DMV Seminar
Band 24

Homotopy Theory and Models

Based on lectures held at a DMV Seminar in Blaubeuren
by H.J. Baues, S. Halperin and J.-M. Lemaire

Marc Aubry

Birkhäuser Verlag
Basel · Boston · Berlin

Author:

Marc Aubry
Laboratoire de Mathématiques
Faculté des Sciences
Université de Nice
Parc Valrose, BP 71
F-06108 Nice Cedex 2
France

A CIP catalogue record for this book is available from the Library of Congress, Washington D.C., USA

Deutsche Bibliothek Cataloging-in-Publication Data

Aubry, Marc:
Homotopy theory and models : based on lectures held at a
DMV seminar in Blaubeuren by H. J. Baues, S. Halperin and
J.-M. Lemaire / Marc Aubry. - Basel ; Boston ; Berlin :
Birkhäuser, 1995
  (DMV-Seminar ; Bd. 24)
  ISBN-13:978-3-7643-5185-4   e-ISBN-13:978-3-0348-9086-1
  DOI: 10.1007/978-3-0348-9086-1

NE: Deutsche Mathematiker-Vereinigung: DMV-Seminar

© 1995 Birkhäuser Verlag, P.O. Box 133, CH-4010 Basel, Switzerland
Camera-ready copy prepared by the author
Printed on acid-free paper produced from chlorine-free pulp
Cover design: Heinz Hiltbrunner, Basel

ISBN-13:978-3-7643-5185-4

9 8 7 6 5 4 3 2 1

Table of contents

Preface

In keeping with the general aim of the "D.M.V.-Seminar" series, this book is principally a report on a group of lectures held at Blaubeuren by Professors H. J. Baues, S. Halperin and J.-M. Lemaire, from October 30 to November 7, 1988. These lectures were devoted to providing an introduction to the theory of models in algebraic homotopy.

The three lecturers acted in concert to produce a coherent exposition of the theory. Commencing from a common starting point, each of them then proceeded naturally to his own subject of research. The reader who is already familiar with their scientific work will certainly give the lecturers their due.

Having been asked by the speakers to take on the responsibility of writing down the notes, it seemed to me that the material elucidated in the short span of fifteen hours was too dense to appear, unedited, in book form. Some amplification was necessary.

Of course I submitted to them the final version of this book, which received their approval. I thank them for this token of confidence. I am also grateful to all three for their help and advice in writing this book. I am particularly indebted to J.-M. Lemaire who was indeed very often consulted.

For basic notions (in particular those concerning homotopy groups, CW-complexes, (co)homology and homological algebra) the reader is advised to refer to the fundamental books written by E. H. Spanier [47], R. M. Switzer [49] and G. Whitehead [52].

Of course there are many very good books and papers dealing with algebraic models. So I never hesitated to give a reference when the present volume failed to provide a detailed proof. On the other hand, I think that until now, no general guide has been available to lead the reader from the basic definitions in algebraic topology to the main results of the theory of models of homotopy types. I hope that this book will satisfy this need.

In conclusion, the lecturers (and this writer) would be happy if this book were to incite non (thus far!)-topologists to read further publications about model theory in algebraic topology. This book also should be an easy way for specialists to quickly locate a result and the references which explain it completely.

Introduction

Algebraic topology attempts to characterize topological spaces and continuous maps by means of algebraic invariants. As J. H. C. Whitehead said at the International Congress of Mathematicians (1950):

> In homotopy theory, spaces are classified in terms of homotopy classes of maps, rather than individual maps of one space on another. Thus, using the word category in the sense of S. Eilenberg and Saunders Mac Lane, a homotopy category of spaces is one in which the objects are topological spaces and the "mappings" are not individual maps but homotopy classes of ordinary maps. The equivalences are the classes with two-sided inverses, and two spaces are of the same homotopy type if and only if they are related by such an equivalence. The ultimate object of algebraic homotopy is to construct a purely algebraic theory, which is equivalent to homotopy theory in the same sort of way that "analytic" is equivalent to "pure" projective geometry.

This book is an introduction to the theory of models in algebraic homotopy. It leads the beginner from basic definitions to specific fields of research. It was never intended to be a complete and self-contained treatise. On the contrary, it is a condensed handbook which should be an introduction to the various models: essentially, the rational models of Sullivan [48] and Quillen [44]. Finally, we sketch an approach to the elaborated results of Baues ([7] and [9]) which extend the "certain exact sequence" of J. H. C. Whitehead [51] and provide in some sense "partial" models over the integrals.

Chapter 1 describes the foundations of algebraic topology; for any space X two basic invariants are recalled:
- the homotopy groups $\pi_n(X) = [S^n, X]$, i.e. homotopy classes of maps from the n-sphere to X;
- the cohomology groups $H^n(X; \mathbb{Z})$.

With respect to these invariants, two kinds of connected spaces play an essential role:
- spaces S such that $H^n(S; \mathbb{Z}) = 0$ except for $n = 0$ and $n = i$ (such as the homology i-sphere);
- spaces K such that $\pi_n(K) = 0$ except for $n = 0$ and $n = i$ (Eilenberg-Mac Lane spaces).

These spaces are "elementary bricks" of iterated constructions of spaces reflecting homotopy (or cohomology) groups, namely the homology (or Postnikov) decomposition.

Chapter 2 develops these constructions in some detail.

The categorical formalism is used as often as possible; in particular the notion of category of cofibration (*chapter 3*) unifies many cases; by this abstract theory one general proof replaces many particular ones.

Various algebraic examples of cofibration categories are described in *chapter 4*.

Homotopy groups are (unfortunately?) endowed with a very rich structure; so rich that there is probably no hope ever to calculate it.

Thus the general problem of finding algebraic models is too difficult. We must restrict ourselves to easier ones.

A first step towards simplification is given by a classical result of Serre which tells us that modulo torsion the homotopy groups of the spheres are quite fair: $\pi_n(S^i) \otimes \mathbb{Q} = \mathbb{Q}$ for $i = 0$ or n if n is odd and for $i = 0$, n and $2n - 1$ if n is even and zero otherwise.

In *chapter 5* we associate to each space X its rationalisation $X_{\mathbb{Q}}(\pi_*(X_{\mathbb{Q}}) = \pi_*(X) \otimes \mathbb{Q})$. We construct two algebraic models for a (simply-connected) space X:
  – one reflecting the Postnikov decomposition of $X_{\mathbb{Q}}$: the Sullivan model is a graded commutative differential algebra the homotopy of which is the rational cohomology of X;
  – the other reflecting the CW-complex decomposition of $X_{\mathbb{Q}}$: the Quillen model is a graded differential Lie algebra the homotopy of which is the rational homotopy of X.

One should remark that from chapter 2 to 5 we develop very parallel theories; it is an important heuristic device called Eckmann-Hilton duality [17].

Chapter 6 elucidates the essential procedure in constructing a CW-complex: the addition of one cell. A similar procedure is examined in algebraic cofibration categories; this leads to the notion of inertia.

Chapter 7 and *8* point out an essential dichotomy: elliptic rational spaces on one side, non-elliptic ones on the other side. With the help of Sullivan models many properties of each class are stated.

Finally, another way to simplify the richness of the homotopy groups of spheres is to restrict ourselves to low-dimensional CW-complexes: then $\pi_*(S^i)$ is known for small i. In fact they are easier to model. J. H. C. Whitehead [51] gave the first results and H. J. Baues [9] extended them widely. *Chapter 9* traces the main lines of this theory.

Chapter 1
Basic Homotopy Theory

The aim of algebraic topology is to classify topological sets and continuous maps with the help of algebraic invariants. Actually most of these invariants define functors; these functors cannot distinguish between two objects which can be "deduced" from each other by a continuous deformation.

So, from the very beginning, it is necessary to introduce the appropriate categories and the notion of homotopy which forms the basis of algebraic topology. This categorical language allows to avoid many repetitions. It will be systematized and constantly used in the description of Sullivan and Quillen theories.

§1 Homotopy

1.1 Definition. Top will denote the category whose objects are topological spaces and whose morphisms are continuous maps. Objects (resp. morphisms) in **Top** will usually simply be called spaces (resp. maps).

Remark. Given spaces X and Y, the set **Top**(X, Y) is equipped with a topology, called the *compact-open* topology, which is generated by all the (open) sets of the form

$$N_{K,U} = \{f \in \mathbf{Top}(X, Y); f(K) \subset U\}, \quad K \subset X \text{ compact, } U \subset Y \text{ open.}$$

Actually, in this book, we are concerned with algebraic tools which are not good enough to characterize objects and morphisms in the category **Top** completely. The best they can do is to distinguish them up to homotopy. This is the notion we introduce in the next paragraphs.

1.2. Let X, Y be spaces and $f, g \in \mathbf{Top}(X, Y)$.

1.2.1 Definition. Let I denote the interval $[0, 1] \subset \mathbb{R}$. We say that f is *homotopic* to g, or that f and g are *homotopic* maps, if and only if there is a map

$$F: X \times I \longrightarrow Y \text{ such that}$$

$$F(x, 0) = f(x) \text{ and } F(x, 1) = g(x).$$

$X \times I$ is called the *cylinder* on X and F is called a *homotopy* between f and g.

Alternatively one can define homotopy in the following "dual" way:

1.2.2 "Dual" definition. We say that f is *homotopic* to g, or that f and g are *homotopic* maps, if and only if there is a map

$$F: X \longrightarrow \mathbf{Top}(I, Y) \text{ such that}$$

$\mathrm{ev}_0 \circ F = f$ and $\mathrm{ev}_1 \circ F = g$, where ev_i is the *evaluation map* at $i \in I$.

The space **Top**(I, Y), also denoted by Y^I, is the space of paths in Y, also called the *path space* on Y.

The equivalence of the two definitions is a consequence of the following theorem, often called the exponential law.

1.3 Theorem. If Y is locally compact and X, Y are Hausdorff, then the canonical map

$$\mathbf{Top}(X \times Y, Z) \to \mathbf{Top}(X, \mathbf{Top}(Y, Z))$$

is a homeomorphism for any space Z.

Proof. See [13].   □

We shall write $f \simeq g$ when the maps f, g are homotopic.

1.4 Proposition. Homotopy is an equivalence relation, compatible with composition of maps, that is:

$$f \simeq g \quad \text{implies} \quad k \circ f \circ h \simeq k \circ g \circ h$$

for any maps h and k for which composition is defined.

1.5 (Corollary and) Definition. There is a category whose objects are spaces and whose morphisms are homotopy classes of maps of **Top**. This category is denoted by $\mathbf{Top}/_\sim$ and we write $[X, Y]$ for $\mathbf{Top}/_\sim(X, Y)$.

A *homotopy equivalence* is an isomorphism in $\mathbf{Top}/_\sim$; two spaces X and Y have the same *homotopy type* if and only if there is a homotopy equivalence $f \colon X \longrightarrow Y$. A space is *contractible* if and only if it is homotopy equivalent to a point.

1.6. We can now formulate some basic problems in homotopy theory:
i)   Describe homotopy types with a given property.
ii)   Given two spaces, describe the set $[X, Y]$.
      (Examples: If X is a compact space, then $\mathrm{Vect}^{\mathbf{R}}(X) = [X, BO(n)]$ and $H^n(X; G) = [X, K(G, n)]$.)
iii) Study extension problems

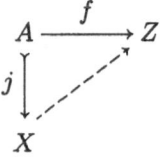

or lifting problems

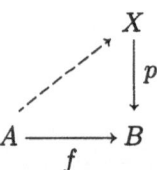

Clearly extension and lifting problems are basic ones in any category. In order that they make sense in the homotopy category, one has to make sure that the existence of a solution only depends on the homotopy class of f. This leads to the notions of fibration and cofibration, which we shall now discuss.

§2 Cofibrations and fibrations

Actually, we shall work with spaces with a distinguished point; a pair (X, x_0) of a space X and a point x_0 in it is called a *pointed space*. Maps between pointed spaces are *pointed maps*, i.e. they preserve base points:

$$f \colon (X, x_0) \longrightarrow (Y, y_0) \text{ is pointed if } f(x_0) = y_0.$$

The definitions of §1.2 can be adapted to the category of pointed spaces, namely:
- A *homotopy* $F \colon X \times I \to Y$ is *pointed* if for all $t \in I$, $F_t \colon X \to Y$ is a pointed map. A pointed homotopy can be viewed as a map $IX \to Y$, where $IX = X \times I / \{x_0\} \times I$ is the *pointed cylinder*.
- The path space Y^I is pointed by the constant map, so that a pointed homotopy can be viewed dually as a pointed map $X \to Y^I$.

Pointed homotopy again is an equivalence relation. The set of pointed homotopy classes is denoted by $[(X, x_0), (Y, y_0)]$ or $[X, Y]^*$.

The exponential law also has a translation into the language of pointed spaces. First denote by $(Y, B)^{(X, A)}$ the subset of Y^X whose elements are the maps f such that $f(A) \subset f(B)$. If (X, x_0), (Y, y_0) are pointed spaces, let us denote by $X \vee Y$ the subspace $X \times y_0 \cup x_0 \times Y$ of $X \times Y$. This space is called the *wedge* of X and Y. Let (Z, z_0) be another pointed space and f_0 be the constant map $f_0 \colon Y \to Z$, $f_0(y) = z_0$. Then the exponential law reads as follows:
If Y is compact and X, Y are Hausdorff, then the exponential map

$$(Z, z_o)^{(X \times Y, X \vee Y)} \to ((Z, z_0)^{(Y, y_0)}, f_0)^{(X, x_0)}$$

induces a homeomorphism for any space Z ([52]).

The correct notion of homotopy groups uses the category of pointed spaces.

From now on, we assume that all spaces are pointed spaces and that all maps and homotopies are also pointed.

2.1 Definition. A map $j \colon A \to X$ is a *cofibration* if any homotopy with source A extends to a homotopy with source X. More precisely, let us consider the following commutative diagram of solid arrows.

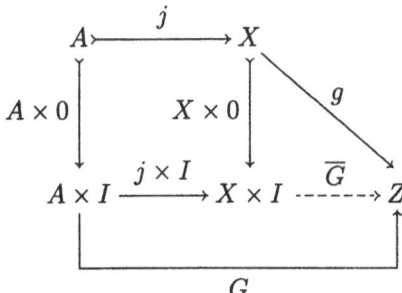

If there is a map $\overline{G}$ (symbolized by the dotted arrow) such that the diagram still commutes, we say that j has the *homotopy extension property* (h.e.p.) with respect to the map g. Then j is a *cofibration* if it has the h.e.p. with respect to any map g.

In diagrams we shall denote cofibrations by $\hookrightarrow$ or $\rightarrowtail$.

Alternatively, using the dual definition of homotopy the h.e.p. means that the diagram

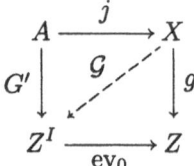

can be filled in by the homotopy $\mathcal{G}$.

Reverting arrows, we readily obtain the dual notion of fibration, i.e. a map with the *homotopy lifting property* (h.l.p.).

More precisely, a map $p: X \to A$ is called a *fibration* if any commutative diagram of solid arrows

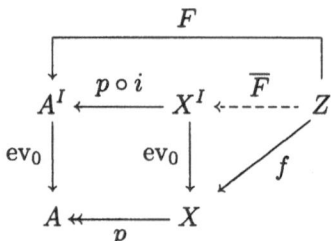

can be filled in by a map $\overline{F}$ (symbolized by the dotted arrow) so that the resulting diagram still commutes. In diagrams we shall denote fibrations by $\twoheadrightarrow$.

As for cofibrations there is a "dual" definition of a fibration: p is a fibration if any commutative diagram

$$
\begin{array}{ccc}
Z \times \{0\} = Z & \xrightarrow{\ f\ } & X \\
\Big\downarrow & \overset{\mathcal{F}}{\nearrow} & \Big\downarrow{\scriptstyle p} \\
Z \times I & \xrightarrow[\ F'\ ]{} & A
\end{array}
$$

can be filled in by a homotopy $\mathcal{F}$; we say that j has the *homotopy lifting property* with respect to the map f.

To characterize cofibrations we need to consider a special case of a push-out. We shall recall the notion of push-out below in §2.7. For a map $j: A \to X$ we denote by $X \cup_A A \times I$ the space which is obtained from the disjoint union of X and $A \times I$ by identifying $j(a) \in X$ with $(a, 0) \in A \times I$ for every $a \in A$.

2.2 Proposition. A map $j: A \rightarrow X$ is a cofibration if and only if $X \cup_A A \times I$ is a retract of $X \times I$.

Proof.

i)   Suppose first that $j: A \hookrightarrow X$ is a cofibration. By definition 2.1, the diagram

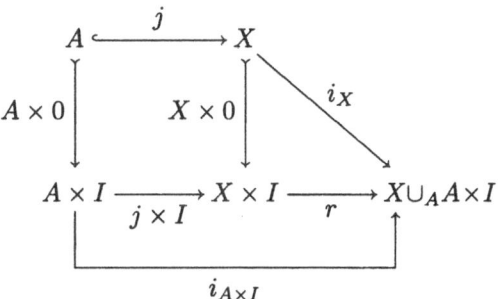

can be completed by the map $r: X \times I \longrightarrow X \cup_A A \times I$ such that

$$r \circ (X \times 0) = i_X \text{ and } r \circ (j \times I) = i_{A \times I},$$

where i_X (resp. $i_{A \times I}$) are the canonical inclusions of X (resp. $A \times I$) into the push-out $X \cup_A A \times I$. Then for the inclusion $i: X \cup_A A \times I \longrightarrow X \times I$, $r \circ i$ is the identity of $X \cup_A A \times I$, i.e. r is a retraction of i.

ii)   Conversely assume the existence of a retraction $r: X \times I \longrightarrow X \cup_A A \times I$. Then we consider the diagram

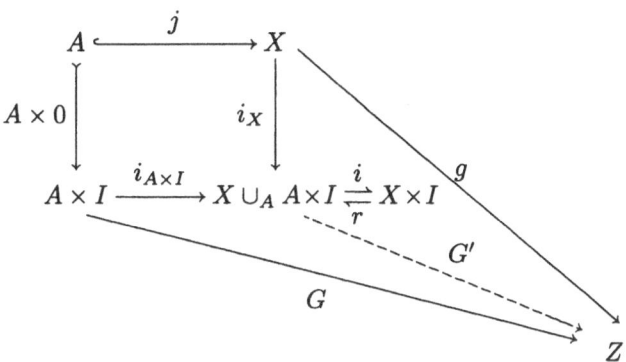

where G' is defined by the push-out diagram (see §2.8) for $X \cup_A A \times I$. One immediately checks:

$$i \circ \bar{j} = j \times I \text{ and } i \circ \overline{A \times 0} = X \times 0.$$

Then $\overline{G} = G' \circ r$ meets the requirements.   □

2.3 Definition. A pointed space (X, x_0) is said to be *well-pointed* if $x_0 \rightarrowtail X$ is a cofibration.

For every space X there is a cofibration $X \rightarrowtail CX$ where CX is a contractible space:

2.4 Definition and proposition. The *cone* of a space X is defined by $X \times I / X \times \{1\}$ and is denoted by $C^{\#} X$. The inclusion $X \rightarrowtail X$ is a cofibration. $C^{\#} X$ is contractible.

The *pointed cone* of a pointed space (X, x_0) is defined by

$$X \times I / (X \times \{1\} \cup \{x_0\} \times I)$$

and is denoted by CX. CX is contractible.

If X is well-pointed, then the inclusion $X \rightarrowtail CX$ is a cofibration.

Dually, for every space Y there is a fibration $W \twoheadrightarrow Y$ where W is a contractible space:

2.5 Definition and proposition. $Y^I \overset{\mathrm{ev}_{0,1}}{\to} Y \times Y$ is a fibration, where $\mathrm{ev}_{0,1}$ is the *evaluation map* at 0 and 1, defined by $\mathrm{ev}_{0,1}(f) = (f(0), f(1))$. The inverse image of $y_0 \times Y$ is called the *track space* relative to y_0 and is denoted by W_{y_0}. The space W_{y_0} is contractible.

2.6 Fundamental property of the path space and the cylinder. The definition of the homotopy relation is based on the spaces Y^I or $Y \times I$. Note that those spaces appear in the following factorizations of the diagonal and of the codiagonal map respectively:

$$
\begin{array}{ccc}
& \Delta & \\
Y \xrightarrow{\ c\ } Y^I \xrightarrow{\ \mathrm{ev}_{0,1}\ } Y \times Y
\end{array}
$$

where c associates to $y \in Y$ the constant path at y, $\mathrm{ev}_{0,1}$ is defined as in 2.5., and

$$
\begin{array}{ccc}
& \nabla & \\
X \amalg X \xrightarrow{\ i\ } X \times I \xrightarrow{\ p\ } X
\end{array}
$$

where $X \amalg X$ denotes the disjoint union of two copies of X, i is the identification of the first (resp. second) copy of X with $X \times \{0\}$ (resp. $X \times \{1\}$) in $X \times I$ and p is the projection on the first factor.

Moreover, c and i are cofibrations, $\mathrm{ev}_{0,1}$ and p are fibrations.

As we said, we want to work in the category of pointed topological spaces. In the first factorization (through the path space) all spaces and maps are canonically pointed if Y is.

This is not the case for the disjoint union $X \amalg X$. However the factorization remains valid if we replace $X \amalg X$ by the *wedge* $X \vee X$ (recall: the wedge (X, x_0) and (Y, y_0) is defined by $X \vee Y = (X, x_0) \amalg (Y, y_0)/x_0 \sim y_0)$ and the cylinder $X \times I$ by the pointed cylinder $IX = X \times I/\{x_0\} \times I$.

More generally, up to a pointed homotopy equivalence, every map can be replaced by a fibration or by a cofibration.

2.7 Proposition ("factorization lemma"). For any map $f : X \longrightarrow Y$ there exist factorizations

$$X \rightarrowtail Z \xrightarrow{\sim} Y$$

and

$$X \xrightarrow{\sim} E \twoheadrightarrow Y$$

which satisfy the symbolized properties.

Examples. The cone, resp. the track space, fit into the following diagrams:

Proof of 2.7 and **definitions.** We can choose for Z the *mapping cylinder* of f defined as $Z_f = X \times I \cup_{(x,1)\sim f(x)} Y$ and then $X \longrightarrow Z$ is the identification with $X \times \{0\} \subset Z_f$ and $Z \longrightarrow Y$ is the map $f \circ p_1 \cup Y : X \times I \cup Y \longrightarrow Y$ (where p_1 is the projection on the first factor $X \times I \longrightarrow X$). (Remark: as in §2.6 our construction is not canonically pointed; in the category of pointed spaces we define the *pointed mapping cylinder* by $Z_f = X \times I \underset{(x,1)\sim f(x)}{\vee} Y$.)

Dually, we can choose for E the *"mapping track"* (which is canonically pointed)

$$E_f = \{(x, \lambda); \lambda(1) = f(x)\} \subset X \times Y^I.$$

Then $X \longrightarrow E$ is defined by $x \mapsto (x, f(x))$
and $E \longrightarrow Y$ by $(x, \lambda) \mapsto \lambda(1)$.

The maps above clearly yield factorizations of f. For the proof that they satisfy the required properties see [47] for instance. □

Cocartesian (resp. cartesian) diagrams behave well with respect to cofibrations (resp. fibrations). This is an essential property of a cofibration (resp. fibration) category (cf. chapter 3). Because of its importance we give a little more details about push-outs.

2.8 Push-outs. The following definition is valid in any category **C**.

2.8.1 Definition. We say that the square $ABCD$ of the diagram is *cocartesian*, or that (D, e, j) realizes a *push-out* of i, k, if the square is commutative and if for any pair of maps $f: B \longrightarrow Z$ and $g: C \longrightarrow Z$ with $f \circ i = g \circ k$, there is a unique map $D \longrightarrow Z$ which makes all subdiagrams commute. If (D, e, j) is a push-out we also call D an *amalgamated sum* of C and B and write $D = C \cup_A B$.

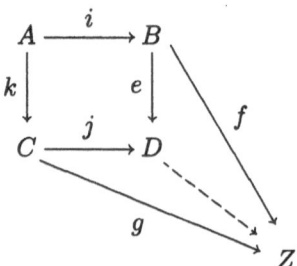

2.8.2 Proposition. Let $\mathbf{C} = \mathbf{Top}$. If i is a cofibration, then j also is a cofibration.

Proof. We use the characteristic property of cofibrations given by proposition 2.2.
    We know that there is a retraction $r: B \times I \longrightarrow (A \times I) \cup_A B$ of the map $(A \times I) \cup_A B \longrightarrow B \times I$.
    Furthermore, we have $D \times I = (C \cup_A B) \times I = (C \times I) \cup_{A \times I} (B \times I)$. Consider the map $\rho := \begin{pmatrix} k \times I & * \\ * & e \end{pmatrix} \circ r: B \times I \longrightarrow (A \times I) \cup_A B \longrightarrow (C \times I) \cup_C D.$ [1] Let us denote the canonical map $C \times I \longrightarrow (C \times I) \cup_C D$ by κ. Then one checks that the map

$$R: C \times I \cup_{A \times I} B \times I \longrightarrow (C \times I) \cup_C (C \cup_A B)$$

defined by $R|_{C \times I} = \kappa$ and $R|_{B \times I} = \rho$ is a retraction of $(C \times I) \cup_C D \longrightarrow D \times I$. By proposition 2.2 $j: C \longrightarrow D$ is a cofibration. □

Warning: The homotopy type of D may change when one replaces i or k by a homotopic map. Therefore amalgamated sums in **Top** in general do not yield a well-defined object in the homotopy category. However, one has:

2.8.3 Proposition. If i is a cofibration, then the homotopy type of D only depends on the homotopy class of k.

Proof. Left to the reader. □

---

(1) We use matricial notation for maps between push-outs in analogy to the notation for maps between direct sums in linear algebra.

2.8.4 Example. Let $\eta\colon S^3 \longrightarrow S^2$ be the Hopf map. Then the complex projective plane $\mathbb{C} P^2 = S^2 \cup e^4$ is the push-out defined by the following diagram:

$$
\begin{array}{ccc}
S^3 & \xrightarrow{\ \ \eta\ \ } & S^2 = \mathbb{C} P^1 \\[2pt]
\Big\downarrow{\scriptstyle k} & & \Big\downarrow{\scriptstyle e} \\[2pt]
e^4 & \longrightarrow & \mathbb{C} P^2
\end{array}
$$

One says that $\mathbb{C} P^2$ is obtained from S^2 by attaching a 4-cell e^4 along the Hopf map $S^3 = \partial e^4 \longrightarrow S^2$. More generally, spaces which can be obtained by iteration of this procedure are called *CW-complexes*.

The reader will observe that $\mathbb{C} P^2$ is the homotopy push-out of the diagram

$$
\begin{array}{ccc}
S^3 & \xrightarrow{\ \ \eta\ \ } & S^2 \\[2pt]
\Big\downarrow & & \\[2pt]
* & &
\end{array}
$$

2.8.5 Definition. Let A, B, C, D be a commutative square in **Top**; let

$$
A \xrightarrow{\ \ i'\ \ } B' \xrightarrow[\alpha]{\ \ \sim\ \ } B
$$

be a factorization of i and let D' be the push-out of i' and k. If the natural map $D' \longrightarrow D$ is a homotopy equivalence, one says that the square is *homotopy cartesian*; one also says that D is the *homotopy push-out* of B and C over A. One can check that this property does not depend on the factorization of i. Indeed, if this property holds for a given factorization of i, it holds for every factorization of i or k.

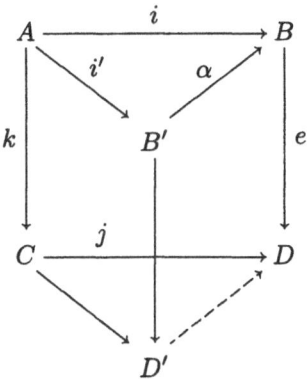

Remark. Push-outs in the strict sense do not exist in general in the homotopy category **Top** $/\sim$ [28]. Therefore homotopy push-out is the relevant notion.

2.9 Cofibre sequence. Important remark: To define an action of the fundamental group on higher homotopy groups (cf. [8] II §5) spaces have to be *well-pointed*, i.e. the inclusion of the base point must be a cofibration.

Once for all, we shall work in the category **Top*** of well-pointed spaces and maps.

2.9.1 Definition. The *mapping cone* or *cofibre* of $f\colon X \to Y$ is the push-out of the following diagram:

$$
\begin{array}{ccc}
X & \xrightarrow{\ f\ } & Y \\
{\scriptstyle k}\downarrow & & \downarrow{\scriptstyle e} \\
CX & \longrightarrow & C_f = Y \cup_f CX
\end{array}
$$

If $Y = *$, C_f is called the *suspension* of X and is denoted by ΣX.

Any sequence $X \xrightarrow{f} Y \xrightarrow{g} Z$ such that

$$
\begin{array}{ccc}
X & \xrightarrow{\ f\ } & Y \\
\downarrow & & \downarrow{\scriptstyle e} \\
* & \longrightarrow & Z
\end{array}
$$

is a homotopy push-out is called a *cofibre sequence*.

2.9.2 Definition and proposition. Collapsing the inclusion of Y in C_f defines the suspension of X; now $Y \longrightarrow C_f \longrightarrow \Sigma X$ is also a cofibre sequence. Continuing this process, one checks that $C_f \longrightarrow \Sigma X \longrightarrow \Sigma Y$ is a cofibre sequence and that the suspension of a cofibre sequence is a cofibre sequence.

Finally we obtain the *long cofibre sequence*

$$
X \xrightarrow{f} Y \longrightarrow C_f \longrightarrow \Sigma X \longrightarrow \Sigma Y \longrightarrow \Sigma C_f \longrightarrow
$$

where each three-term sequence is a cofibre sequence.

Proof. See the original article [5] and [42] or [34]. See also [49] and [52]. $\square$

2.9.3 Proposition. If

$$
X \xrightarrow{\ j\ } Y \xrightarrow{\ r\ } Z
$$

is a cofibre sequence, then

$$
[X, W]^* \xleftarrow{\ j^*\ } [Y, W]^* \xleftarrow{\ r^*\ } [Z, W]^*
$$

is an exact sequence of pointed sets.

Proof. It is an easy consequence of Definition 2.9.1 of the cofibre. See [42] for details. □

The long cofibre sequence is endowed with an algebraic structure. We first recall that any suspension has a *cogroup*[2] structure in the homotopy category:

2.9.4. Proposition. For any space X, the suspension ΣX is a cogroup in **Top***/~; furthermore $[\Sigma X, W]^*$ is a group.

Proof. See the references of footnote 2. We only recall that a cogroup structure on ΣX can be depicted as follows

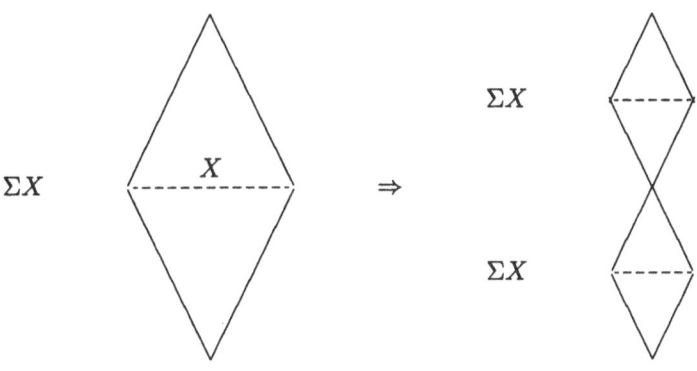

□

One can check that the map $C_f \longrightarrow C_f \vee \Sigma X$ symbolized by the following sketch

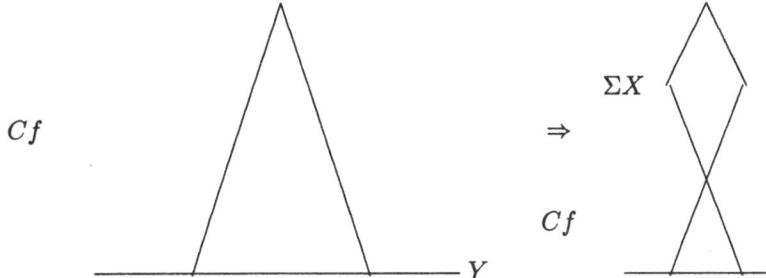

defines an action of the group $[\Sigma X, W]^*$ on $[C_f, W]^*$, denoted by $*$:

---

(2) For a definition of a *cogroup* see [45] and [32].

$$[C_f \vee \Sigma X, W]^* \longrightarrow [C_f, W]^*$$

$$\sim \downarrow \qquad \qquad \nearrow$$

$$[C_f, W]^* \times [\Sigma X, W]^*$$

Now we are in a position to state the following assertion:

2.9.5 Proposition. The long exact cofibre homotopy sequence is:

$$[X, W]^* \xleftarrow{\; j^* \;} [Y, W]^* \xleftarrow{\; r^* \;} [Z, W]^* \longleftarrow [\Sigma X, W]^* \longleftarrow [\Sigma Y, W]^* \longleftarrow \quad \cdots$$

where

i)   The first two arrows from the left form an exact sequence of sets with the following additional property:
For $f, f' \colon Z \to W$ we have $f \circ r \sim f' \circ r$ iff there is a map $g \colon \Sigma X \to W$ such that $g * f = f$.

ii)   The next arrow is given by $g \mapsto 0 * g$ and describes the action of the group $[\Sigma X, W]^*$ on the trivial map $0 \in [Z, W]^*$.

iii)   With the exception of these three arrows, this an exact sequence of groups.

iv)   Moreover, if W is a group in **Top**$^*/_\sim$ (for instance a topological group, or a loop space), the whole sequence is an exact sequence of groups (of abelian groups at the stage $[\Sigma\Sigma X, W]^* \leftarrow [\Sigma\Sigma Y, W]^* \leftarrow [\Sigma\Sigma Z, W]^* \leftarrow \dots$).

Proof. [34]. $\square$

Example. One has $S^3 \times S^3 = (S^3 \vee S^3) \cup e^6$. Now S^3 admits a topological group structure (SU(2) or Sp(1)). So $[S^3 \times S^3, S^3]$ is a group. The long cofibre sequence begins with

$$S^3 \vee S^3 \longrightarrow S^3 \times S^3 \longrightarrow S^6 \longrightarrow S^4 \vee S^4 \longrightarrow \Sigma(S^3 \times S^3).$$

So

$$[S^3 \vee S^3, S^3] \leftarrow [S^3 \times S^3, S^3] \leftarrow [S^6, S^3] \leftarrow [S^4 \vee S^4, S^3] \leftarrow [\Sigma(S^3 \times S^3), S^3].$$

Now $[S^3 \vee S^3, S^3] \leftarrow [S^3 \times S^3, S^3]$ is surjective; in fact, if m is the multiplication of S^3, then $f, g \mapsto m \circ (f \times g)$ is a section of pointed sets (not of groups because S^3 is not homotopy commutative).

On the other hand, $[S^4 \vee S^4, S^3] \leftarrow [\Sigma(S^3 \times S^3), S^3]$ is also surjective, because $\Sigma(S^3 \times S^3) \sim S^4 \vee S^4 \vee S^6$.

So the long exact sequence breaks up into

$$0 \leftarrow \mathbb{Z}^2 \leftarrow [S^3 \times S^3, S^3] \leftarrow \mathbb{Z}/12\,\mathbb{Z} \leftarrow 0.$$

Such a group extension is classified by an element in $H^2(\mathbb{Z}^2; \mathbb{Z}/12\mathbb{Z}) = \mathbb{Z}/12\mathbb{Z}$. The above extension is in fact a generator of $\mathbb{Z}/12\mathbb{Z}$. This can be seen using the following facts

a) $[S^3 \times S^3, S^3]$ is a nilpotent group of nilpotence 2 ([52]).

b) The commutator $p_1 p_2 p_1^{-1} p_2^{-1}$ of the two projections $p_i \colon S^3 \times S^3 \longrightarrow S^3$, $i = 1, 2$, represents a homotopy class e in $[S^3 \times S^3, S^3]$; clearly e is in fact in the kernel of the map $\mathbb{Z}^2 \longleftarrow [S^3 \times S^3, S^3]$ and therefore it can be identified with an element of $\mathbb{Z}/12\mathbb{Z}$.

Moreover this element e is a generator of $\mathbb{Z}/12\mathbb{Z}$ (cf. [45]).

Chapter 2
Homology and Homotopy Decomposition
of Simply Connected Spaces

In this chapter we show that each simply connected homotopy type can be represented either by a "homology decomposition" or by a "homotopy decomposition" (the latter is more commonly known as a Postnikov decomposition or a Postnikov "tower").

We have already seen instances of such a "duality" (called Eckmann-Hilton duality, cf. [17]), namely fibrations versus cofibrations. Indeed Eckmann-Hilton duality can be viewed as a categorical duality, i.e. arising from "reversing the arrows".

However, in the case of topological spaces, this duality fails to be strict and can merely be used as a heuristic principle.

Recall that categorical duality is based on the concept of an opposite category $\mathbf{C}^{op}$ associated to a category $\mathbf{C}$. The category $\mathbf{C}^{op}$ has the same objects as $\mathbf{C}$, the morphism sets of $\mathbf{C}^{op}$ are defined by

$$\mathbf{C}^{op}(X,Y) = \mathbf{C}(Y,X)$$

with the correspondance

$$(Y \xleftarrow{f^{op}} X) \leftarrow (Y \xrightarrow{f} X)$$

where composition is defined in $\mathbf{C}^{op}$ by $f^{op} \circ g^{op} = (g \circ f)^{op}$.

Each categorical concept like sum or push-out gives us a dual concept by reversing the arrows in the definitions. For example we have the dual concepts

$$\text{sum} \longleftrightarrow \text{product}$$

$$\text{push-out} \longleftrightarrow \text{pull-back}$$

More precisely we have

$$(\text{sum in } \mathbf{C}) = (\text{product in } \mathbf{C}^{op})$$

$$(\text{push-out in } \mathbf{C}) = (\text{pull-back in } \mathbf{C}^{op}).$$

Let us recall the two following statements from chapter 1 (§2.6.):

i) Any map $f: X \longrightarrow Y$ factors as a cofibration followed by a homotopy equivalence.

ii) Any map $g: Y \longrightarrow X$ factors as a homotopy equivalence followed by a fibration.

Clearly these two statements correspond to each other under reversing the arrows and exchanging "fibration" and "cofibration". This is a basic example of "Eckmann-Hilton duality".

We begin with more examples of Eckmann-Hilton dual concepts.

§1 Eckmann-Hilton duality

(From now on we shall always use the following notation: $(X, *)$ is a well-pointed space and t is an element of $I = [0,1] \subset \mathbb{R}$.)

1.1 "Definition". In the category $\mathbf{Top}^*$ of well-pointed spaces the following notions are Eckmann-Hilton "dual" [17].

Cofibre	Fibre
Cylinder	*Path space*
$X \longrightarrow IX = I \times X / I \times *$	$X \xleftarrow{p_t} PX = (X^I, *)$
$x \xmapsto{i_t} (x, 0)$	$p_t(\sigma) = \sigma(t)$
$i_t(x) = (t, x)$	
homotopy extension property: cofibration in **Top**	homotopy lifting property: fibration in **Top**
The suspension $\Sigma \colon \mathbf{Top}^* \to \mathbf{Top}^*$ is defined by	The loop space $\Omega \colon \mathbf{Top}^* \to \mathbf{Top}^*$ is defined by

$$
\begin{array}{ccc}
* & \longleftarrow & X \\
{\scriptstyle k}\downarrow & \text{p.o.} & \downarrow{\scriptstyle i_1} \\
CX \longleftarrow IX & \xleftarrow{\; i_0 \;} & X \\
\downarrow & \text{p.o.} & \downarrow \\
\Sigma X & \longleftarrow & *
\end{array}
\qquad\qquad
\begin{array}{ccc}
* & \longleftarrow & X \\
{\scriptstyle k}\uparrow & \text{p.b.} & \uparrow{\scriptstyle i_1} \\
WX \longrightarrow PX & \xrightarrow{\; p_0 \;} & X \\
\uparrow & \text{p.b.} & \downarrow \\
\Omega X & \longleftarrow & *
\end{array}
$$

or simply by the homotopy push-out $\qquad\qquad$ or simply by the homotopy pull-back

$$
\begin{array}{ccc}
X & \longrightarrow & * \\
\downarrow & & \downarrow \\
* & \longrightarrow & \Sigma X
\end{array}
\qquad\qquad\qquad
\begin{array}{ccc}
\Omega X & \dashrightarrow & * \\
\downarrow & & \downarrow \\
* & \longrightarrow & X
\end{array}
$$

$$[\Sigma X, Y]^* \simeq [X, \Omega Y]^*$$

(generalized homotopy groups)

The map $\nabla: \Sigma X \longrightarrow \Sigma X \vee \Sigma X$ symbolized by the following picture

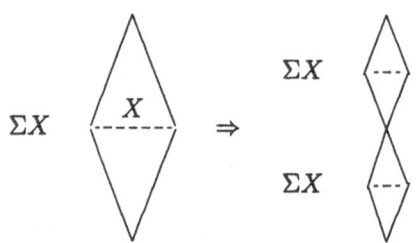

defines a co-H-space structure on ΣX.

The addition of paths

$$m: \Omega X \times \Omega X \longrightarrow \Omega X$$

defines a H-space structure on ΩX.

Or using the second definition of the suspension the following diagram defines ∇

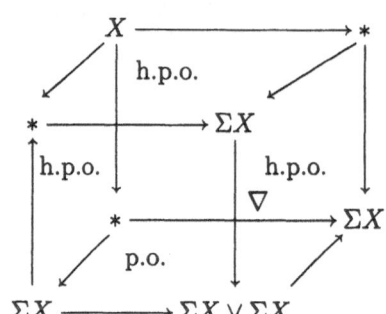

Or using the second definition of the loop space the following diagram defines m

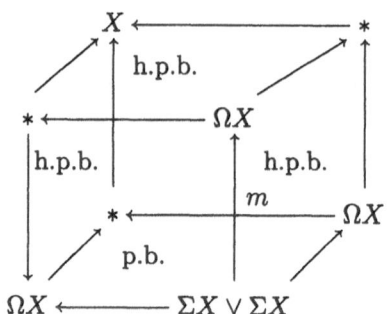

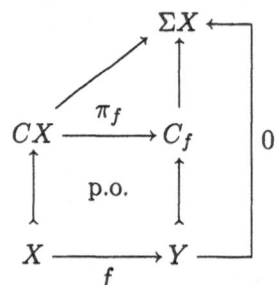

$Y \longrightarrow C_f \longrightarrow \Sigma X$ is a *principal cofibration* with cofibre ΣX and co-classifying map f.

C_f is the *(homotopy) cofibre* or *mapping cone*.

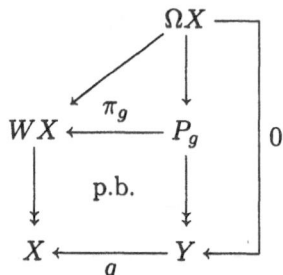

$\Omega X \longrightarrow P_g \longrightarrow Y$ is a *principal fibration* with fibre ΩX and classifying map g.

P_g is the *(homotopy) fibre* of g.

For completeness we recall the existence and properties of the cofibre sequence (see chapter 1 §§2.9.2 & 2.9.5) and of its Eckmann-Hilton dual.

Given $f: X \longrightarrow Y$, we obtain the cofibre sequence of f:

$$cof(f) = \left\{ X \xrightarrow{f} Y \rightarrowtail C_f \to \Sigma X \xrightarrow{\Sigma f} \cdots \to \Sigma^2 X \xrightarrow{\Sigma^2 f} \cdots \right\}$$

and the fibre sequence of f:

$$fib(g) = \left\{ X \xleftarrow{g} Y \leftarrow P_g \leftarrow \Omega X \xleftarrow{\Omega g} \cdots \leftarrow \Omega^2 X \xleftarrow{\Omega^2 g} \cdots \right\}.$$

1.2 Remark. Proposition 2.9.3 states that $[cof(f), U]^*$ is an exact sequence (of sets everywhere and compatible with certain group structures according to the range). A similar property holds for $[U, fib(g)]^*$.

At this point we want to introduce relative homotopy groups ([49], Chapter 3). Let $g: Y \hookrightarrow X$ be a cofibration (of pointed spaces). We define the $(n+1)$-th *relative homotopy group* of the pair (X, Y) by

$$\pi_{n+1}(X, Y) = \pi_n E_g = [S^n, E_g]^*.$$

where E_g denotes as usual the *mapping track* of g.

Homotopy can be carried over to (pointed) pairs of spaces as for single spaces. In particular the *exponential law* yields the identification:

$$\pi_{n+1}(X, Y) = [(CS^n, S^n); (X, Y)]$$

(CS^n is the cone over the n-sphere; it is homeomorphic to the $n + 1$-ball, often denoted by D^{n+1}).

A glance at the definition immediately yields:

1.3 Proposition. There is a long exact sequence of homotopy groups of pairs

$$\longrightarrow \pi_{n+1}(X, Y) \longrightarrow \pi_n(Y) \longrightarrow \pi_n(X) \longrightarrow \pi_n(X, Y) \longrightarrow \cdots$$

□

We now introduce the "building blocks" of homology and homotopy decompositions.

1.4 Moore spaces.

1.4.1 Definition. Let A be an abelian group. A *Moore space* of type (A, n) is a connected, simply connected space M such that

$$H_i M(A, n) = \begin{cases} 0 & \text{if } i \neq 0 \text{ or } n \\ \mathbb{Z} & \text{if } i = 0 \\ A & \text{if } i = n \end{cases}$$

Here is a construction of such a space $M(A, n)$. Choose a presentation of the group A:

$$\bigoplus_J \mathbb{Z} \overset{d}{\rightarrowtail} \bigoplus_I \mathbb{Z} \to A \; .$$

Let δ be the (unique!) map $\bigvee_J S^n \overset{\delta}{\longrightarrow} \bigvee_I S^n$ such that $H_n(\delta) = d$. We put $M(A, n) = C_\delta$. The homotopy type of C_δ is well defined (cf. infra), and meets the requirements.

1.4.2 Proposition. Any two Moore spaces of type (A, n) have the same homotopy type. Therefore by slight abuse of language we shall speak of *the* Moore space of type (A, n) and denote it by $M(A, n)$.

Proof. Let M be a Moore space of type (A, n). By the Hurewicz theorem (cf. [47], Chapter 7), we have

$$\pi_n(M) = A, \quad \text{and} \quad \pi_i(M) = 0 \quad \text{if} \quad i < n.$$

Choose a presentation of A as above

$$\bigoplus_J \mathbb{Z} \rightarrowtail \bigoplus_I \mathbb{Z} \to A.$$

There exists a map

$$\bigvee_I S^n \overset{g}{\longrightarrow} M$$

such that $\pi_n(g) = p$. Since $p \circ d = 0$, the composite $g \circ \delta$ is null-homotopic. Therefore there is a map

$$g' : M(A, n) \longrightarrow M$$

which induces the identity $A = A$ on π_n.

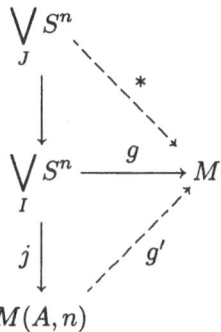

By naturality of the Hurewicz map, g' also induces the identity on H_n and therefore on H_*. By the Whitehead theorem (cf. [47] chapter 7), g' is a homotopy equivalence.   □

1.4.3 Proposition. If $f: A \longrightarrow B$ is a morphism of abelian groups, one can define a map $M(f,n): M(A,n) \longrightarrow M(B,n)$ up to homotopy, such that the correspondance $M(.,n): \mathbf{Ab} \longrightarrow \mathbf{Top}/_\sim$ is functorial.

Proof. Apply the same ideas as in proof of 1.4.2.   □

1.5 Eilenberg-Mac Lane spaces.

1.5.1 Definition. Let A be an abelian group. An *Eilenberg-Mac Lane space* of type (A,n) is a connected space K such that

$$\pi_i K = \begin{cases} 0 & \text{if } i \neq 0 \text{ or } n \\ \mathbb{Z} & \text{if } i = 0 \\ A & \text{if } i = n . \end{cases}$$

Given (A,n), we sketch two constructions of such a space:

(1) Construction of $K(A,n)$ by killing higher homotopy groups.

We start with $M(A,n)$, whose first non-trivial homotopy group is $\pi_n M(A,n) = A$ by Hurewicz and then kill higher homotopy groups. We call $K(A,n)$ the obtained space. The proof relies on the following steps.

1.5.2 "Killing π_n" lemma. Let X be a CW-complex. Then there is a CW-complex Y and an inclusion $i: X \subset Y$ such that $\pi_n Y = 0$ and $i_*: \pi_i X \xrightarrow{\sim} \pi_i Y$ is an isomorphism for $j < n$.

Proof. Let $f_\alpha: S^n \longrightarrow X$ be generators of $\pi_n X$. Let $Y = C_f = X \cup e^{n+1} \cup \cdots \cup e^{n+1}$, where $f = \vee_\alpha f_\alpha: S^n \longrightarrow X$. Consider the exact homotopy sequence of the pair (Y, X):

$$\pi_{n+1}(Y, X) \twoheadrightarrow \pi_n X \twoheadrightarrow \pi_n Y \longrightarrow \pi_n(Y, X).$$

We have $\pi_n(Y, X) = 0$ by homotopy approximation (cf. [49] chapter 6, theorem 6.10). The map $\pi_{n+1}(Y, X) \longrightarrow \pi_n X$ is a surjection. Indeed, by construction of Y, to every map $f_\alpha: S^n \longrightarrow X$ corresponds a so-called characteristic map $(D^{n+1}, S^n) \longrightarrow (Y, X)$ which identifies D^{n+1} with e^{n+1} and ∂D^{n+1} with $f_\alpha(S^n)$.

From the exactness of the sequence we deduce that $\pi_n Y = 0$. □

1.5.3 Corollary 1. (Killing π_j, $j > n$). Let X be a CW-complex. Then there is a CW-complex $X^{[n]}$ and an inclusion $i: X \subset X^{[n]}$ such that $\pi_j X^{[n]} = 0$ for $j > n$ and $i_j: \pi_j X \simeq \pi_j X^{[n]}$ for $j \leq n$.

Note that $M(A, n)^{[n]}$ will thus be an Eilenberg-Mac Lane space of type (A, n).

Proof. Applying the lemma above inductively, we obtain a sequence of inclusions

$$X = Y_0 \subset Y_1 \subset Y_2 \subset \cdots \subset Y_{i-1} \subset Y_i \subset \cdots$$

where $\pi_j Y_i = \pi_j Y_{j+1}$ if $j \leq n + i$ and $\pi_{n+i} Y_{i+1} = 0$.

We set $X^{[n]} = \lim Y_r$ and use the fact that $\pi_i \lim Y_r = \lim(\pi_i Y_r)$ if (Y_{i+1}, Y_i) are relative CW-complexes for all i (cf. [49] chapter 6). □

(2) Construction of $K(A, n)$ using the infinite symmetric product (cf. [16]).

1.5.4 Definition. For any space pointed let X^n be the n-fold product of X. Let the symmetric group Σ_n act on X by permuting factors. The quotient space $SP_n(X) = X^n/\Sigma_n$ is called the *n-fold symmetric product*.

Note that any of the inclusions $X^n \subset X^{n+1}$ given by setting one coordinate equal to the base point induces the same inclusion $j_n: SPX^n \subset SPX^{n+1}$.

$$X^n \xrightarrow{\;* \times X^n\;} X^{n+1}$$

$$SP_n \xrightarrow[\;j_n\;]{} SP_{n+1}$$

If X is well-pointed the latter is a cofibration and the *infinite symmetric product* of X is defined by $SP_\infty(X) = \lim SP_n(X)$.

1.5.5 Theorem (Dold-Thom). If X is a 1-connected space, there is a natural isomorphism

$$H_i(X) = \pi_i SP_\infty(X).$$

Moreover the Hurewicz morphism is induced by the inclusion

$$X = SP_1(X) \subset SP_\infty(X).$$

Proof. Cf. [16]. □

1.5.6 Corollary. $SP_\infty(M(A,n))$ is an Eilenberg-Mac Lane space of type (A,n).

As for Moore spaces, an Eilenberg-Mac-Lane space is well-defined up to homotopy by its type. We shall therefore feel free to speak of the Eilenberg-Mac Lane space $K(A,n)$.

We leave the proof to the reader. □

1.5.7 Proposition. If $f: A \longrightarrow B$ is a morphism of abelian groups, one can define a map $K(f,n): K(A,n) \longrightarrow K(B,n)$ up to homotopy, such that the correspondance $K(.,n): \mathbf{Ab} \longrightarrow \mathbf{Top}/_\sim$ is functorial.

Proof. Prove that the construction of the infinite symmetric product is functorial and use 1.4.3. □

1.6 Examples.

i) Moore spaces:
- $M(\mathbb{Z},k) = S^k$, the ordinary sphere.
- We define the rational sphere as the space $M(\mathbb{Q},k) = S^k_\mathbb{Q}$,
- $P_n = S^1 \cup_n e^2$, the pseudo-projective plane ($P_2 = \mathbb{R}P_2$). One easily verifies that
$$M(\mathbb{Z}/n\mathbb{Z}, k) = \Sigma^{k-1} P_n;$$
    it is also called the torsion sphere.

ii) Eilenberg-Mac Lane spaces:
- $K(\mathbb{Z},1) = S^1$
- $K(\mathbb{Z}/2\mathbb{Z},1) = \mathbb{R}P_\infty$ (its universal cover is $S^\infty = \varinjlim_n S^n$)
- $K(\mathbb{Z},2) = \mathbb{C}P_\infty$ (Sketch of proof: use the homotopy exact sequence of $S^1 \longrightarrow S^{2n+1} \longrightarrow \mathbb{C}P(n)$ and pass to the limit)
- for n odd, $K(\mathbb{Q},n) = (S^n)_\mathbb{Q}$ (Proof: using the Serre spectral sequence one can easily show that $H^*(K(\mathbb{Q},n)) = H^*((S^n)_\mathbb{Q})$).

We now list some easy properties of Moore and Eilenberg-Mac Lane spaces.

1.7 Proposition.
$$M(A \oplus B, n) = M(A,n) \vee M(B,n)$$
$$K(A \oplus B, n) = K(A,n) \wedge K(B,n)$$
$$\Sigma M(A,n) = M(A, n+1)$$
$$\Omega K(A,n) = K(A, n-1), \text{ for } n > 1.$$

For any extension of abelian groups $0 \longrightarrow A' \longrightarrow A \longrightarrow A'' \longrightarrow 0$, one has a fibre sequence
$$K(A',n) \longrightarrow K(A,n) \longrightarrow K(A'',n)$$
and a cofibre sequence
$$M(A',n) \longrightarrow M(A,n) \longrightarrow M(A'',n)$$

Proof. Exercise.   □

1.8 Definition. Let X be a space, A an abelian group and n any integer; $[M(A,n),X]^*$ is called the *n-th homotopy group of X with coefficients in A* and denoted by $\pi_n(A;X)$. Dually, $[X,K(A,n)]^*$ is called the *n-th cohomology of X with coefficients in A* and denoted by $H_n(A;X)$.

   The homotopy exact sequence associated to

$$K(A',n) \longrightarrow K(A,n) \longrightarrow K(A'',n)$$

gives rise to the so-called *long exact sequence for cohomology with coefficients.*

   Dually, there is a long exact sequence for homotopy with coefficients ([34]).

   We are now ready to proceed to the core of this chapter.

§2 Homology and homotopy decompositions

2.1 Definition. Let X be a 1-connected space, then a homology decomposition is defined by the following data:

For every integer $n \geq 2$, spaces $X_{(n)}$ and maps $\alpha_n \colon X_{(n)} \longrightarrow X$ and $k'_{n+1} \colon M(H_{n+1}(X), n) \longrightarrow X_{(n)}$ such that

1) $X_{(2)} = M(H_2(X), 2)$;
2) X_{n+1} is the cofibre of k'_{n+1};
3) $H_j \alpha_n$ is an isomorphism for $j \leq n$.

   A homotopy decomposition is defined by the following data:

For every integer $n \geq 2$, spaces X_n and maps $\beta_n \colon X \longrightarrow X_n$ and $k_{n+1} \colon X_n \longrightarrow K(\pi_{n+1}(X), n+2)$ such that

1) $X_2 = K(\pi_2(X), 2)$;
2) X_{n+1} is the pull-back of the principal fibration defined on $K(\pi_{n+1}(X), n+2)$;
3) $\pi_j \beta_n$ is an isomorphism for $j \leq n$.

2.2 Theorem. Each 1-connected homotopy type $\{X\}$ admits both a homotopy decomposition and a homology decomposition.

The properties of homology and homotopy decompositions can be summarized by the following commutative diagrams.

Homology decomposition   Homotopy decomposition

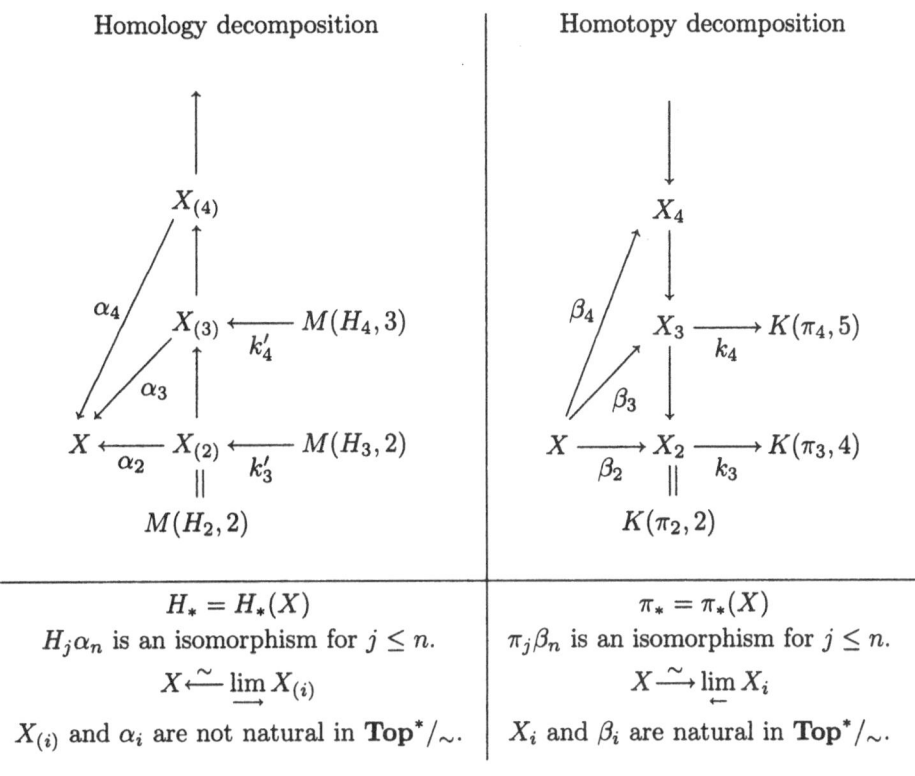

$H_* = H_*(X)$	$\pi_* = \pi_*(X)$
$H_j\alpha_n$ is an isomorphism for $j \leq n$.	$\pi_j\beta_n$ is an isomorphism for $j \leq n$.
$X \xleftarrow{\sim} \varinjlim X_{(i)}$	$X \xrightarrow{\sim} \varprojlim X_i$
$X_{(i)}$ and α_i are not natural in **Top**$^*/_\sim$.	X_i and β_i are natural in **Top**$^*/_\sim$.

(Duality breaks down at this point.)

The proof will make use of the following classical theorems of Hurewicz and Whitehead.

2.3 Definition. A map $f: X \longrightarrow Y$ between two path-connected spaces is called a *weak homotopy equivalence* if and only if $\pi_*(f)$ is an isomorphism.

2.4 Whitehead theorem. With the above notations: $H_*(f)$ is an isomorphism iff $\pi_*(f)$ is.

2.5 Theorem. If X and Y are CW-complexes, a weak homotopy equivalence $f: X \longrightarrow Y$ is a homotopy equivalence.

Proof. For both theorems see [47]. ☐

Remark. We point out again that although $X \longrightarrow X_n$ is a functor from **Top**$/_\sim$ into itself, the homotopy type of $X_{(n)}$ is not determined by X (cf. [11]). However, this is the case if X is a rational space (see chapter 5).

2.6 Proof of the existence of a homology decomposition. It uses the following four intermediate results:

2.6.1 Lemma (*principal reduction*). Let $\xi: A \longrightarrow B$ and $g: B \longrightarrow X$ be two maps such that $g \circ \xi$ is null-homotopic, and let $H: A \times I \longrightarrow X$ be the null-homotopy. Let $G_g: X \rightarrowtail C_g$ be the cofibre of g, and G be the canonical null-homotopy $G: B \times I \longrightarrow C_g$, i.e. the composite $B \times I \to CB \rightarrowtail C_g$.

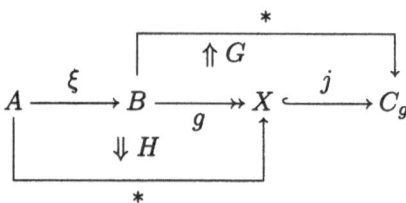

Let us consider the following diagram:

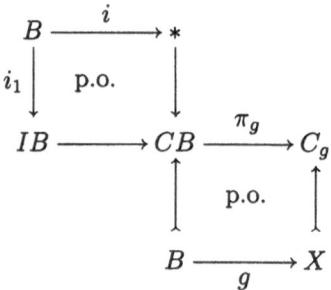

Let $f: \Sigma A \longrightarrow C_g$ be defined by $f = -jH + G\pi_\xi$ and let $\xi_g: C_\xi \longrightarrow X$ be the map $\xi_g = (g, H)$ defined by the push-out.

Then the *double principal cofibration*

$$X \subset C_g \subset C_f$$

is equivalent to the principal cofibration $X \subset C_{\xi_g}$, that is, there is a homotopy equivalence

$$C_{\xi_g} \sim C_f \text{ under } X.$$

Proof. It is a special case of [6], 3.5.2. and follows from the diagram

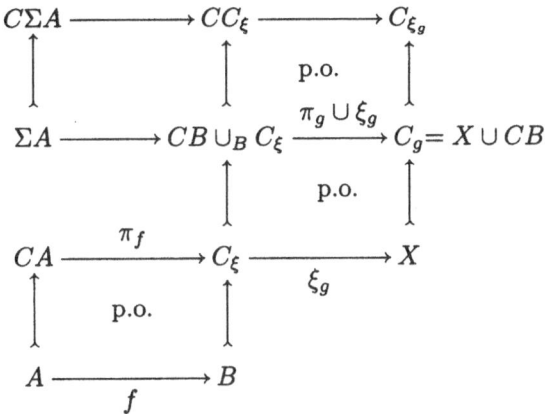

2.6.2 Lemma. Let us define the *generalized homotopy groups* ([47]):

$$\pi_n^A(X) = [\Sigma^n A, X] = \pi_n^A(X, *)$$

$$\pi_n^A(X,Y) = [(\Sigma^{n-1}CA, \Sigma^{n-1}A), (X,Y)]$$

and consider the following diagram

$$\pi_1^A(CB, B)$$

$$\downarrow (\pi_g)_*$$

$$f \in \pi_1^A(C_g, *) \longrightarrow \pi_1^A(C_g, X)$$

where j is induced by the inclusion $* \subset X$ and B is 1-connected.

If $j(f) \in \text{image}(\pi_g)_*$, then there exists a map $\xi \colon A \longrightarrow B$ and a homotopy $H \colon A \times I \longrightarrow X$ with $f = -H + G$ as above.

Proof. See [6] p 221.   □

2.6.3 Lemma. If A and B are wedges of n-spheres, $n \geq 2$, then $\pi_1(X) = 0$.

Proof. S^n can be considered as the CW-complex obtained by attaching an n-cell to a point; then $\pi_i(S^n, *) = 0$, $i \leq n - 1$, by homotopy approximation (cf. [49] chapter 6, theorem 6.10). Now apply the van Kampen theorem ([52], chapter II).□

2.6.4 Corollary. Let X be a CW-complex equipped with a cellular decomposition such that $X^1 = *$, where X^n denotes the n-skeleton. Let Z_n be the set of n-cells of $X(n \geq 2)$, and let

$$\xi \colon \bigvee_{Z_{n+1}} S^{n+1} \longrightarrow \bigvee_{Z_n} S^n = X^n / X^{n-1}$$

be the obvious projection (called the *cellular boundary* of X). Then there is a map

$$\xi_g \colon C_\xi \longrightarrow X^n$$

with $X^{n+2} \simeq C_{\xi_g}$ rel X^n.

Proof. Let $g: \bigvee_{Z_n} S^n \longrightarrow X^n$ be the attaching map of the n-cells and let $f: \bigvee_{Z_{n+1}} S^{n+1} \longrightarrow X^{n+1}$ be the attaching map of the $(n+1)$-cells.

Because of lemma 2.6.3 and ([6] theorem 3.4.7)

$$(\pi_g)_*: \pi_1^A(CB, B) \longrightarrow \pi_1^A(C_g, X)$$

is surjective. Then lemma 2.5.2 applies. Hence the corollary is an application of the principal reduction 2.5.1.   □

We can now proceed to build up the homology decomposition.

Let $C_n = H_n(X^n, X^{n-1})$ and consider

$$C_{n+2} \xrightarrow{d_{n+1}} C_{n+1} \xrightarrow{d_n} C_{n-1}.$$

As usual set $B_i = d_{i+1}C_{i+1}$,

$$Z_{i+1} = \ker(d_{i+1}: C_{i+1} \longrightarrow C_i) \text{ for every } i > 0.$$

Since B_{n-1} is free there is a map s:

$$H_n \subset C_n/B_n \xleftarrow[\longrightarrow]{s} B_{n-1}.$$

Therefore if $\xi: \bigvee_{Z_{n+1}} S^{n+1} \longrightarrow \bigvee_{Z_n} S^n$ is the map in corollary 2.5.4 we have

$$H_n(C_\xi) = H_n \oplus sB_{n-1}; \quad H_{n+1}(C_\xi) = Z_{n+1}.$$

Then lemma 2.6.5 proves the existence of the homotopy equivalence

$$C_\xi \sim M(H_n, n) \vee M(sB_{n-1}, n) \vee M(Z_{n+1}, n+1).$$

2.6.5 Lemma. If X is $(n-1)$-connected and $(n+1)$-dimensional, then

$$X \sim M(H_n X, n) \vee M(H_{n+1} X, n+1).$$

Proof. First recall that if A is free abelian

$$[M(A, n), X] \xrightarrow{j} \operatorname{Hom}(A, \pi_n X).$$

Now the Hurewicz map $h_i: \pi_i X \longrightarrow H_i X$ is an isomorphism for $i = n$ and an epimorphism for $i = n+1$ (for X is $(n-1)$-connected). As H_{n+1} is free abelian, we can choose a section t of h_{n+1}. Let φ_n (resp. φ_{n+1}) be a representant of

$$j^{-1}(H_n X) \in [M(H_n X, n), X] \text{ (resp. of } j^{-1}t(H_{n+1}X) \in [M(H_{n+1}X, n+1), X]).$$

Consider

$$\varphi: \varphi_n \vee \varphi_{n+1}: M(H_n X, n) \vee M(H_{n+1}X, n+1) \longrightarrow X.$$

By construction $H_i(\varphi)$ is an isomorphism for every i. As the spaces are 1-connected the Whitehead theorem (cf. §2.3.) asserts that φ is a homotopy equivalence.   □

We now define X_n as the mapping cone of the composite

$$M(sB_{n-1}, n) \longrightarrow C_\xi \xrightarrow{\xi_g} X^n$$

and we define k'_{n+1} as the composite

$$k'_{n+1} \colon M(H_n, n) \longrightarrow C_\xi \xrightarrow{\xi_g} X^n \subset X_n.$$

□

We have now described a homology decomposition of X; let us point out that it depends on a cellular decomposition of X and on the choice of the section s in $(*)$.

2.6.6 Remark. If $H_n X$ is a free abelian group for all n, then the homology decomposition is a CW-decomposition with a minimal number of cells since

$$M\left(\bigoplus_I \mathbb{Z}, n\right) = \bigvee_I S^n.$$

Examples of such a minimal decomposition are given by

$$\mathbb{C}P_\infty = S^2 \cup e^4 \cup e^6 \cup \dots$$

$$S^n \times S^m = (S^n \vee S^m) \cup_\omega e^{n+m}$$

$$\Omega S^{n+1} = S^n \cup_{[i_n, i_n]} e^{2n} \cup e^{3n} \cup \dots \text{ (cf [36])}.$$

Here ω is the *Whitehead product* map (cf [6] chapter 0 and [52] chapter X). Recall that the Whitehead product $\pi_n(X) \times \pi_m(X) \longrightarrow \pi_{n+m-1}(X)$ is defined by means of ω as

$$[\alpha, \beta] = \omega^*(\alpha, \beta) \colon S^{n+m-1} \xrightarrow{\omega} S^n \vee S^m \xrightarrow{(\alpha, \beta)} X.$$

Also recall that the Samelson product

$$\pi_{n-1}(\Omega X) \times \pi_{m-1}(\Omega X) \longrightarrow \pi_{n+m-2}(\Omega X)$$

is defined by the commutative diagramm

$$
\begin{array}{ccc}
(\alpha, \beta) \in \pi_n(X) \times \pi_m(X) & \xrightarrow{\;[,\,]\;} & \pi_{n+m-1}(X) \\[4pt]
{\scriptstyle \tau \times \tau} \downarrow \simeq & & \simeq \downarrow {\scriptstyle (-1)^{n-1}\tau} \\[4pt]
(x, y) \in \pi_{n-1}(\Omega X) \times \pi_{m-1}(\Omega X) & \xrightarrow{\;[,\,]\;} & \pi_{n+m-2}(\Omega X)
\end{array}
$$

where the vertical arrows are given by the adjointness relation

$$\pi_n X = [SS^{n-1}, X] \overset{\tau}{\cong} [S^{n-1}, \Omega X] = \pi_{n-1}\Omega X.$$

2.6.7. Lemma. The Samelson product gives $\pi_* \Omega X$, $* \geq 1$, the structure of a graded Lie algebra, that is

$$[x, y] = -(-1)^{|x| \cdot |y|} [x, y]$$
$$[x, [y, z]] = [[x, y], z] + (-1)^{|x| \cdot |y|} \quad [y, [x, z]].$$

Proof. See [52] chapter X.   □

2.7 Proof of the homotopy decomposition. As for the homology decomposition, the construction of the homotopy decomposition will be achieved in several steps.

Let X be a connected CW-complex.

2.7.1 Lemma. There exists a commutative diagram

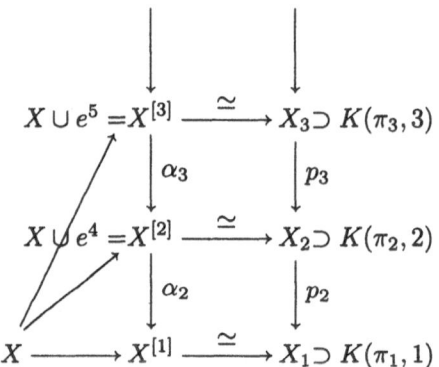

where p_n is a fibration with fibre $K(\pi_n, n)$.

Proof. First, the spaces $X^{[n]}$ in the diagram above are obtained from X by gluing cells of dimension $\geq n + 2$ and verify $\pi_j X^{[n]} = 0$ for $j > n$ and $\pi_j X^{[n]} = \pi_j X$ for $j \leq n$ (apply the procedure of killing homotopy groups, cf. §1.5.2). The map $X \longrightarrow X^{[1]} = K(\pi_1, 1)$ represents the first Postnikov invariant. For constructing α we use the extension lemma:

2.7.2 Lemma (Extension lemma). Let $\pi_n Y = 0$, Y connected, and let $f: X \longrightarrow Y$ be a map. Then, if $\overline{X}$ is obtained from X by attaching m-cells, $m \geq n+2$, there is an extension $\overline{f}: \overline{X} \longrightarrow Y$.

Proof. (Cf. [6] 1.2.2).   □

The space $X^{[n]}$ is obtained from X by adjoining m-cells, $m \geq n+2$, and

$$\pi_n X^{[n-1]} = 0.$$

Then we can recursively factor all the α according to the factorization lemma.

We now wish to prove the existence of the right part of the diagram for the homotopy decomposition (actually the mapping in the Postnikov tower). Suppose we have already built the maps up to stage n as indicated by the following diagram:

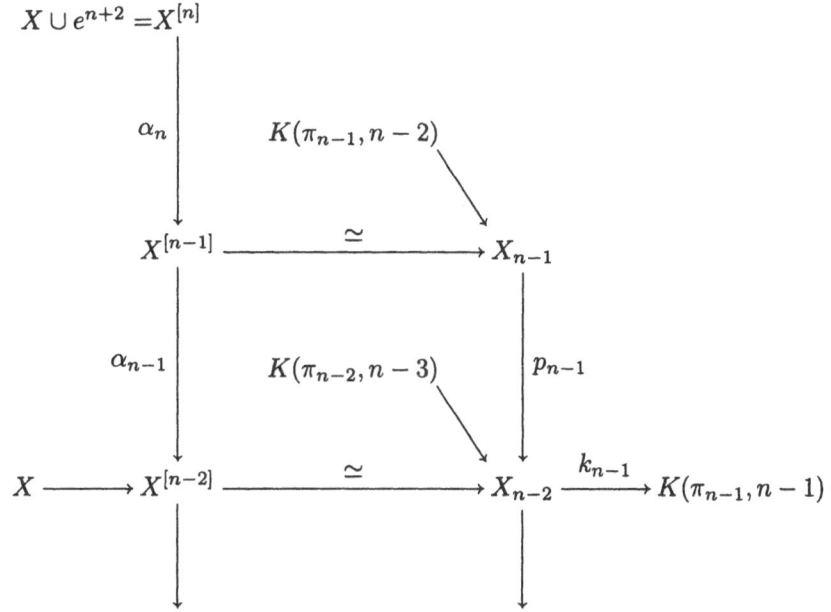

Now we apply the factorization lemma (chapter 1 §2.6) to the composite:

$$f\colon X^{[n]} \xrightarrow{\ \alpha\ } X^{[n-1]} \xrightarrow{\ \simeq\ } X_{n-1}$$

Hence we get a homotopy equivalence $X^{[n]} \xrightarrow{\sim} X_n$ and a fibration $p_n\colon X \xrightarrow{\simeq} X_{n-1}$.

We now want to exhibit that p_n is a principal fibration. The exact homotopy sequence for the fibration p_n proves that the fibre is a $K(\pi_n, n)$ (recall that $\pi_j X^{[n]} = 0$ for $j > n$ and $\pi_j X^{[n]} = \pi_j X$ for $j \le n$). Therefore we use:

2.7.4 Lemma (Classification of $K(A,n)$-fibrations). Let X be a simply connected space. Then any fibration p with fibre $K(A,n)$

$$
\begin{array}{c}
K(A,n) \subset X \\
\ \ \ \ \ \downarrow p \\
X_{n-2} \dashrightarrow^{k} K(A, n+1)
\end{array}
$$

is a principal fibration. The element $k \in H^{n+1}(X, A)$ is the first obstruction for the existence of a cross section of p (cf. [6] 4.3.7) and p is equivalent to the fibration induced by k from the canonical one

$$K(A, n) \longrightarrow PK(A, n) \longrightarrow K(A, n+1).$$

Proof. See [6] 5.2.2.   □

This completes the proof of the homotopy decomposition.   □

§3 Application: Classification of 2-stage spaces

As an application of homology and homotopy decomposition, we indicate how one can describe all homotopy types with only two non-vanishing homotopy (resp. homology) groups.

Let $M(H_n, H_m)$ be the set of all homotopy types $\{X\}$ with $1 < n < m$ and

$$H_i(X) = \begin{cases} H_n & \text{if } i = n \\ H_m & \text{if } i = m \\ 0 & \text{otherwise.} \end{cases}$$

3.1 Theorem. There is a bijection

$$\{f \in \pi_{m-1}(H_m, M(H_n, n)) : H_* f = 0\}/\sim \; \simeq M\{H_n, H_m\}.$$

Proof. To each map $f : M(H_m, m-1) \longrightarrow M(H_n, n)$ associate the mapping cone $\{C_f\}$. The existence of the homology decomposition proves that this correspondance is a surjection which factors through the following equivalence relation:
$f \sim f'$ iff there are homotopy equivalences $\alpha \in \text{Aut}(M(H_n, n))$ and $\beta \in \text{Aut}(M(H_m, m-1))$ with $\beta_* f = \alpha^* f'$.

For $n > 2$ this surjection is also an injection by the theory of principal maps between fiber spaces (See chapter V §7 in [8]). For $n = 2$ the equivalence relation $\sim$ cannot be so simple (it uses "twisted maps") in order to still get a bijection. □

Let $K(\pi_n, \pi_m)$ be the set of all homotopy types $\{Y\}$ with $1 < n < m$ and

$$\pi_i(X) = \begin{cases} \pi_n & \text{if } i = n \\ \pi_m & \text{if } i = m \\ 0 & \text{otherwise.} \end{cases}$$

3.2 Theorem. There is a bijection

$$H^{m+1}(K(\pi_n, n), \pi_m)/\sim \; \simeq M\{H_n, H_m\}.$$

Proof. To each map $g: K(\pi_n, n) \longrightarrow K(\pi_m, m+1)$ associate the mapping fibre $\{P_g\}$. The existence of the homotopy decomposition proves that this correspondance is a surjection which factors through the following equivalence relation:

$g \sim g'$ iff there are homotopy equivalences $\alpha \in \mathrm{Aut}(K(\pi_n, n))$ and $\beta \in \mathrm{Aut}(K(H_m, m+1))$ with $\beta_* g = \alpha^* g'$.

This surjection is also an injection by the theory of principal maps between fiber spaces (see chapter V §10 in [8]). □

Remark. The groups $H^{m+1}(K(\pi_n, n), \pi_m)$ are "in principle" known by the work of Eilenberg-Mac Lane [18] [19] and Cartan [12].

Example. Let $X = \Sigma(\mathbb{R} P_4)$ be the suspension of the real projective 4-space. This is a 1-connected space. It is well-known, that $H_i(\mathbb{R} P_4) = \begin{cases} \mathbb{Z}/2\mathbb{Z} & \text{for } i = 1, 3 \\ 0 & \text{otherwise} \end{cases}$

Therefore

$$H_2 = \mathbb{Z}/2\mathbb{Z}$$
$$H_3 = 0$$
$$H_4 = \mathbb{Z}/2\mathbb{Z}$$
$$\cdots$$
$$H_n = 0, \; n > 4.$$

and by theorem 4.1 $X = C_f$ where $f = k_4' \neq 0$.

$$0 \neq k_4' \in \quad \mathrm{Ext}(\mathbb{Z}/2\mathbb{Z}, \mathbb{Z}/4\mathbb{Z}) = \mathbb{Z}/2\mathbb{Z}$$

$$f \in \quad \pi_3(\mathbb{Z}/2\mathbb{Z}, M(\mathbb{Z}/2\mathbb{Z}, 2))$$

$$0 \in \quad \mathrm{Hom}(\mathbb{Z}/2\mathbb{Z}, \mathbb{Z}/4\mathbb{Z}) = \mathbb{Z}/2\mathbb{Z}$$

The right column is a universal coefficient exact sequence (See [41]).

$\Sigma(\mathbb{R} P_4)$ is a homology 2-stage space with $H_2 = H_4 = \mathbb{Z}/2\mathbb{Z}$, $H_i = 0$ for $i \neq 2, 4$. Its homotopy type is entirely determined by k_4'. Furthermore $M(H_2, H_4)$ consists of two spaces up to homotopy: $\Sigma(\mathbb{R} P_4)$ and $M(\mathbb{Z}/2\mathbb{Z}, 2\mathbb{Z}) \vee M(\mathbb{Z}/2\mathbb{Z}, 4)$.

The homotopy decomposition of $\Sigma(\mathbb{R} P_4)$ begins as follows; we list the first homotopy groups

$$\pi_2 = \mathbb{Z}/2\mathbb{Z}$$
$$\pi_3 = \mathbb{Z}/2\mathbb{Z}$$
$$\pi_4 = \mathbb{Z}/4\mathbb{Z}$$
$$\cdots$$
$$\pi_n = ?, \; n > 4.$$

And the first k-invariant k_3 is the generator of the following isomorphic groups:

$$k_3 \in H^4(K(\pi_2, 2), \pi_3)$$

$$\|$$

$$\mathrm{Hom}(H_4(K(\pi_2, 2)), \pi_3)$$

$$\|$$

$$\mathrm{Hom}(\mathbb{Z}/4\mathbb{Z}, \mathbb{Z}/2\mathbb{Z}) = \mathbb{Z}/2\mathbb{Z}$$

since $H_3 K(\pi_2) = 0$. Here we use the universal coefficient theorem.

Chapter 3
Cofibration Categories

We now proceed to set up a suitable abstract framework which contains all ingredients for constructing and describing algebraic models of homotopy theory. We aim at an axiomatization of the minimal properties used to develop a homotopy theory. The category **Top** introduced in chapter 1 is a typical example of a cofibration category; actually, the axiomatization given below originated from the study of this category.

§1 Basic definitions

1.1 Definition. A *cofibration category* $(\mathbf{C}, cof, we)$ is a category $\mathbf{C}$ together with two classes of morphisms cof, we satisfying the axioms C_1, C_2, C_3 and C_4 below. Morphisms in cof (resp. we) are called *cofibrations* (resp. *weak equivalences*).

Morphisms in $\mathbf{C}$ are also called maps in $\mathbf{C}$. We write $i\colon B \subset A$ or $B \rightarrowtail A$ for a cofibration and we denote the restriction of $u\colon A \longrightarrow U$ by $u|_B = u \circ i\colon B \longrightarrow U$. We write $X \xrightarrow{\sim} Y$ for a weak equivalence in $\mathbf{C}$. An isomorphism in $\mathbf{C}$ is denoted by $\cong$. We shall use the same notation for an object as for its identity map, e.g. X.

The axioms for a cofibration category are:

(C_1) Composition axiom: given two maps

$$A \xrightarrow{f} B \xrightarrow{g} C,$$

if any two of the maps f, g and gf are weak equivalences, then so is the third. A composite of cofibrations is again a cofibration. Moreover, the isomorphisms in $\mathbf{C}$ are both weak equivalences and cofibrations.

(C_2) Push-out axiom: for a cofibration $i\colon A \rightarrowtail B$ and a map $f\colon A \longrightarrow Y$, there exists a push-out in $\mathbf{C}$

$$
\begin{array}{ccc}
A & \xrightarrow{\ f\ } & Y \\
\downarrow{\scriptstyle i} & & \downarrow{\scriptstyle \bar{i}} \\
B & \xrightarrow[\ \bar{f}\ ]{} & B \cup_A Y = B \cup_f Y
\end{array}
$$

and $\bar{i}$ is a cofibration. Moreover:
a)   if f is a weak equivalence, so is $\bar{f}$.
b)   if i is a weak equivalence, so is $\bar{i}$.

(C_3) Factorization axiom: given any map $f\colon B \longrightarrow Y$ in $\mathbf{C}$ there is a commutative diagram

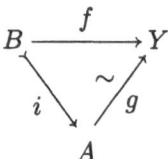

where i is a cofibration and g a weak equivalence.

Before stating the fourth axiom, we introduce the following definition:

1.2 Definition. A map in $\mathbf{C}$ is a *trivial cofibration* if it is both a weak equivalence and a cofibration. An object R in a cofibration category $\mathbf{C}$ will be called a *fibrant model* (or simply *fibrant*) if each trivial cofibration

$$i: R \xrightarrow{\sim} Q \quad \text{in } \mathbf{C}$$

admits a retraction

$$r: Q \longrightarrow R, \ r \circ i = R.$$

(C_4) Axiom on fibrant models: for each object X in $\mathbf{C}$ there is a trivial cofibration $X \xrightarrow{\sim} RX$ where RX is fibrant in $\mathbf{C}$. Such a map $X \xrightarrow{\sim} RX$ is called a *fibrant model* of X.

Remark. The categorical dual of a cofibration category is a *fibration category*. We leave the details to the reader.

1.3 Definition. Assume that $\mathbf{C}$ has an initial object $*$. An object X in $\mathbf{C}$ is said to be $*$-*cofibrant* if $* \rightarrowtail X$ is a cofibration.

Quillen [43] has introduced the notion of a *(closed) model category*. This is a category $\mathbf{M}$ together with classes of morphisms *cof*, *fib*, *we* satisfying certain axioms. We only mention here that the cofibrant objects in $\mathbf{M}$ form a cofibration category, the fibrant objects in $\mathbf{M}$ form a fibration category.

Let $\mathbf{C}_c$ be the full subcategory of $\mathbf{C}$ which consists of $*$-cofibrant objects. One easily checks that $(\mathbf{C}_c, \text{Mor } \mathbf{C} \cap cof, \text{Mor } \mathbf{C}_c \cap we)$ is a cofibration category in which all objects are $*$-cofibrant. We shall sometimes write cofibrant instead of "$*$-cofibrant"; but we point out that the notion "$*$-cofibrant" depends on the existence of an initial object $*$ and is not the dual of the notion "fibrant object" defined in (C_4). Anyway the dual of (C_4) only makes sense in a *fibration category*.

§2 Homotopy in a cofibration category

First we define *cylinder objects* in a cofibration category as follows: let $B \rightarrowtail A$ be a cofibration. By (C_2) we have the push-out diagram

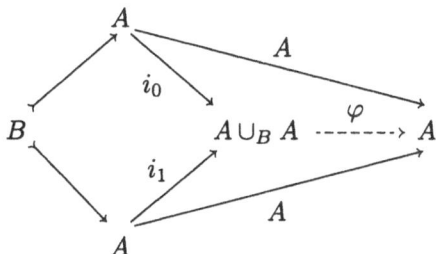

where $\varphi = (A, A)$ is called the *folding map*. By (C_3) φ admits a factorization:

$$A \cup_B A \overset{i}{\rightarrowtail} Z \overset{p}{\underset{\sim}{\longrightarrow}} A .$$

We call (Z, i, p) a *relative cylinder* on $B \rightarrowtail A$ and use the notation $Z = I_B A$. By (C_1), the maps $(i_0, i_1) \colon A \rightarrowtail Z$ are trivial cofibrations since $p \circ i_0 = p \circ i_1 = A$.

2.1 Definition. Two maps $\alpha, \beta \colon A \longrightarrow X$ in a cofibration category are *homotopic relative to B* (or *under B*) and we write $\alpha \simeq \beta$ rel B if there is a commutative diagram

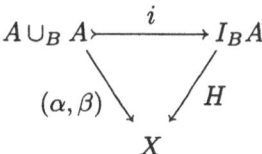

where $I_B A$ is a relative cylinder on $B \rightarrowtail A$. We call H a homotopy from α to β relative to B.

In general, the property of maps $A \longrightarrow X$ "to be homotopic" is not an equivalence relation. But this holds if X is fibrant, cf. lemma 3.7 below.

Fundamental example. The category **Top** defined as in chapter 1, with homotopy equivalences (§1.5 chapter 1) as weak equivalences and cofibrations defined in §2.1 (chapter 1), is a cofibration category in which all objects are fibrant.

In Table 1 we list some examples of cofibration categories. We refer to [8] for a thorough description.

Table 1: Topological examples of cofibration categories

Notation	**Top**	**Top$_{CW}$.**	**CW$^+$**
Objects	topological spaces	connected pointed top. spaces	(X, N_X), X CW-space well-pointed, connected $N_X \subset \pi_1(X)$, perfect normal subgroup
Morphisms	maps	pointed maps	pointed maps $f: X \to Y$ with $f_*(N_X) \subset N_Y$
we	homotopy equivalences	π_*-equivalences = weak homotopy equivalences	$f_*: \pi_1(X)/N_X \cong \pi_1(Y)/N_Y$ such that for any $\pi_1(Y)/N_Y$-module L $f: \hat{H}_*(X, f_*L) \cong \hat{H}_*(Y, L)$
cof	HEP = homotopy extension property	inductive attachments of cells	HEP
fibrant	all objects	all objects	objects of the form $(X, 0)$; the fibrant model of (X, N_X) is the Quillen (+)-construction $(X^+, 0)$

We denote by k_* a generalized homology theory defined on CW-pairs with the limit axiom; that is this theory satisfies: $\lim_{\to} k_* X_\alpha = k_* X$, where X_α is a finite subcomplex. Each such homology theory leads to a cofibration category.

Table 2: k_*-cofibration categories

Notation	**CW$_{k_*}$**	**CW$_{\mathbb{Q}}$**	**CW$_R$**	**CW$_{\mathbb{Z}/p}$**
Objects	CW-spaces	CW-spaces	CW-spaces	CW-spaces
Morphisms	maps	maps	maps	maps
we	k_*-equivalences			
cof	homotopy extension property			
k_*		$H_*(-, \mathbb{Q})$	$H_*(-, R)$, $R \subset \mathbb{Q}$	$H_*(-, \mathbb{Z}/p)$
fibrant	k_*-local spaces	rational spaces	R-local spaces	p-complete spaces

§3 Properties of cofibration categories

Recall that every object X admits a fibrant model for RX, i.e., that there is a map $X \overset{\sim}{\rightarrowtail} RX$ where RX is fibrant in **C**.

3.1 Extension lemma. Let i and f be given as in the diagram

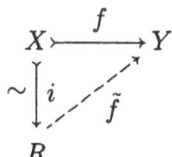

Then, if Y is fibrant, there is an extension $\tilde{f}$ with $\tilde{f} \circ i = f$.

Moreover, two extensions $\tilde{f}$, $\tilde{f}_1$ of f are homotopic relative to X.

Proof. Let P be the push-out of (i, f). Then $Y \longrightarrow P$ is a trivial cofibration (Axiom C_2). Since Y is fibrant by definition there is a retraction $r: P \longrightarrow Y$

$$
\begin{array}{ccc}
X & \overset{f}{\longrightarrow} & Y \\
\sim \downarrow i \quad \text{p.o.} & & \sim \downarrow \bar{i} \quad r \\
R & \underset{\bar{f}}{\longrightarrow} & P
\end{array}
$$

Then $\tilde{f} = r\bar{f}$ is the required extension.

The second part of the lemma follows from the diagram:

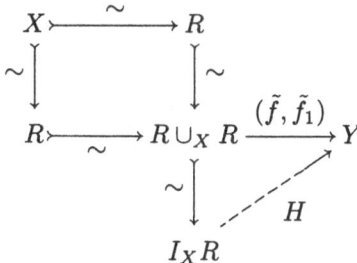

More precisely, axiom C_2 and the definition of the cylinder show that $R \cup_X R \longrightarrow I_X R$ is a trivial cofibration. Now we can apply the first part of the lemma to define a homotopy H. □

3.2 Weak lifting lemma. Any commutative diagram

(∗)

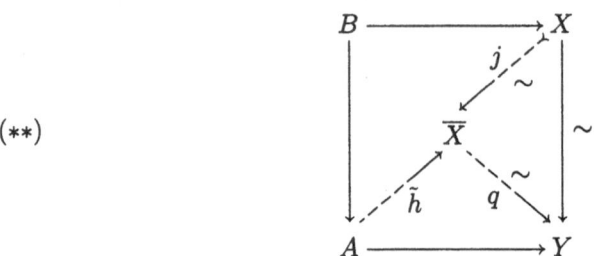

can be embedded into a commutative diagram of the following form:

(∗∗)

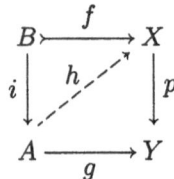

We call the pair $L = (\tilde{h}, j)$ in (2) a *weak lifting* for diagram (∗). The map $\tilde{h}$ is a cofibration provided $B \longrightarrow X$ is a cofibration.

Proof. Apply (C_3) to the map $A \cup_B X \longrightarrow Y$ which is defined by (∗) and use (C_1).
□

3.3 Lifting lemma. Consider the commutative diagram of solid arrows:

a)   If X is fibrant there is a map h for which the upper triangle commutes.
b)   If X and Y are fibrant there is map h for which the upper triangle commutes and for which $p \circ h$ is homotopic to $g \operatorname{rel} B$. We call a map h with these properties a lifting for the diagram.
c)   If X and Y are fibrant, a lifting for the diagram is unique up to homotopy rel B.

Proof. Since X is fibrant there is a retraction $r: \overline{X} \longrightarrow X$ in (**) of 3.2. Let $h = r\tilde{h}$. This proves a).

Now assume that X and Y are fibrant. With the notations of 3.2 we have to prove

$$ph = pr\tilde{h} = qjr\tilde{h} \simeq g \operatorname{rel} B.$$

We consider the following commutative diagram of solid arrows:

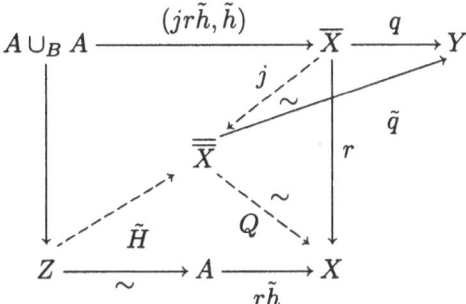

Here Z denotes a cylinder on $B \subset A$. As in 3.2 we obtain the weak lifting $\tilde{H}$ and since Y is fibrant we get the map $\tilde{q}$ by the extension lemma 3.1. Thus $H = \tilde{q} \circ \tilde{H}$ is a homotopy. This prove b). We leave c) to the reader.   □

As an application we now show that a classical theorem of Dold (see [14] and [15]) holds in any cofibration category.

3.4 Theorem of Dold [13]. Consider the commutative diagram

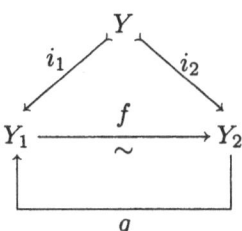

If Y_1, Y_2 are fibrant the map f is a homotopy equivalence rel Y, i.e., there is a map

$$g: Y_2 \longrightarrow Y_1 \text{ such that } g \circ i_2 = i_1 \text{ and } gf \simeq Y_1 \operatorname{rel} Y, \ fg \simeq i_2 \operatorname{rel} Y.$$

Proof. Consider the following diagram of solid arrows:

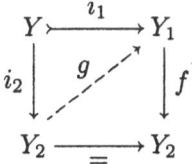

The lifting lemma 3.3 proves the existence of a map $g\colon Y_2 \longrightarrow Y_1$ with $g \circ i_2 = i_1$ (assertion a) and $f \circ g \simeq Y_2$ (assertion b).

Applying assertion c) to the diagram above proves that $g \circ f$ is homotopic to Y_1. □

3.5 Corollary ("Uniqueness" of fibrant models). All fibrant models of a given object are homotopy equivalent under X.

Proof. Let $i_1 \overset{\sim}{\rightarrowtail} RX$ and $i_2\colon X \overset{\sim}{\rightarrowtail} R'X$ be two fibrant models of X. The extension lemma 3.1 yields the existence of a map f such that the following diagram commutes:

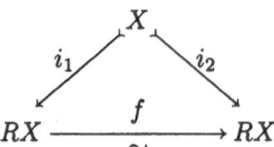

The map f is a homotopy equivalence under X by the theorem of Dold 3.4. □

We conclude this paragraph with an example of a cofibration category where all objects are not fibrant, namely $\mathbf{CW_Q}$ (for a definition see Table 2). Here, for any space X the fibrant model is

$$X \overset{\sim}{\rightarrowtail} RX = X_{(0)},$$

where $X_{(0)}$ is the rationalization of X (cf. chapter 5).

As we have already seen, fibrant objects play a central role in any cofibration category. We now give some examples.

3.6 Corollaries of the lifting lemma.

3.6.1 Corollary. Let $i\colon Y \overset{\sim}{\rightarrowtail} X$ be a cofibration and a weak equivalence between fibrant objects. Then Y is a *deformation retract* of X. That is, there is a retraction

$$r\colon X \longrightarrow Y$$

with $r \circ i = Y$ and with $i \circ r \simeq X \operatorname{rel} Y$.

Proof. Consider the diagram

$$
\begin{array}{ccc}
Y & \overset{Y}{\longrightarrow} & Y \\
{\scriptstyle i}\downarrow{\scriptstyle \sim} & & {\scriptstyle \sim}\downarrow{\scriptstyle i} \\
X & \underset{X}{\longrightarrow} & X
\end{array}
$$

and apply 3.3. □

3.6.2 Corollary. Let Y be fibrant and let

$$B \xrightarrow{i} A_1 \xrightarrow[\sim]{p} Y,$$

$$B \xrightarrow{j} A_2 \xrightarrow[\sim]{q} Y$$

be factorizations (by axiom (C_3)) of a given $f: B \longrightarrow Y$ such that A_1 and A_2 are fibrant. Then up to homotopy rel B there is a unique weak equivalence $\alpha: A_1 \longrightarrow A_2$ with $\alpha \circ i = j$ and $q \circ \alpha \simeq p \operatorname{rel} B$.

Proof. Straightforward. □

3.6.3 Corollary. A retract of a fibrant object is fibrant.

Proof. Let X be a retract of a fibrant Y:

$$X \underset{r}{\overset{i}{\rightleftarrows}} Y \quad \text{with} \quad r \circ i = X.$$

Let $X \overset{\sim}{\underset{\alpha}{\rightarrowtail}} Z$ be a trivial cofibration.

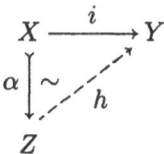

Then by the lifting lemma 3.3 there is a lifting h such that $i = h \circ \alpha$, and $r \circ h$ is the required retraction. □

Finally we mention the fundamental property of sets of homotopy classes.

3.7 Lemma. Let $i: B \rightarrowtail A$ be a cofibration and let U be fibrant, then the homotopy relation $\simeq$ (cf. §2) is an equivalence relation on the set $\{f: A \longrightarrow U \mid fi = u\}$ which is independant of the choice of the cylinder $I_A B$.

Proof. See [8], chapter 2. We write $[A, U]_i^B = [A, U]^U = \{f: A \longrightarrow U \mid fi = u\}/ \simeq$ rel B. $[A, U]_i^* = [A, U]$ if A is cofibrant and $*$ is the initial object. □

§4 Properties of cofibrant models

4.1 Definition. Let $\mathbf{C}$ be a cofibration category with an initial object $*$. By (C_3) the canonical map $* \longrightarrow X$ can be factored as

$$* \rightarrowtail RMX \overset{\sim}{\underset{\alpha}{\rightarrow}} X.$$

We call $MX \overset{\sim}{\rightarrow} X$ a *cofibrant model* of X. We now quote two basic facts on cofibrant models.

4.2 Lemma. If X is fibrant we can choose MX to be fibrant.

Proof. Let $\alpha: MX \xrightarrow{\sim} X$ be a cofibrant model of the fibrant object X and choose (axiom (C_4)) a fibrant model $MX \rightarrowtail RMX$ of X. Then α factors through RMX by α' (cf. extension lemma 3.1). By axiom (C_1) α' is a quasi-isomorphism. Then the factorization $* \rightarrowtail RMX \overset{\sim}{\underset{\alpha}{\rightarrow}} X$ settles the question. □

4.3 Proposition (uniqueness of cofibrant models). Let

$$* \rightarrowtail MX \xrightarrow[p]{\sim} X \qquad \text{and} \qquad * \rightarrowtail M'X \xrightarrow[p']{\sim} X$$

be both cofibrant and fibrant models of the fibrant object X. Then there is a unique quasi-isomorphism α up to homotopy equivalence rel $*$

$$\alpha: MX \xrightarrow{\sim} M'X \quad \text{with} \quad p'\alpha \simeq p.$$

Proof. This is an immediate application of the lifting lemma 3.3. □

4.4 Example of cofibrant model. In $\mathbf{Top_{CW}}$. (for a definition see Table 1) a cofibrant model of a space X is a pointed CW-complex MX together with a weak equivalence

$$* \rightarrowtail MX \xrightarrow{\sim} X.$$

§5 The homotopy category as a localization

5.1 Localization.

5.1.1 Definition. Let $\mathbf{C}$ be an arbitrary category and let S be a subclass of the class of morphisms in $\mathbf{C}$. By the localization of $\mathbf{C}$ with respect to S we mean the category $S^{-1}\mathbf{C}$ together with functor $q: \mathbf{C} \longrightarrow S^{-1}\mathbf{C}$ having the following universal property: for every $s \in S, q(s)$ is an isomorphism; given any functor $F: \mathbf{C} \longrightarrow \mathbf{B}$ such that $F(s)$ is an isomorphism for all $s \in S$, there is an unique functor $\theta: S^{-1}\mathbf{C} \longrightarrow \mathbf{B}$ with $\theta q = F$. In a suitable environment the category $S^{-1}\mathbf{C}$ exists (see [29]). Let $\mathrm{Ho}(\mathbf{C})$ be the localization of $\mathbf{C}$ with respect to the given class of weak equivalences in $\mathbf{C}$. We call it the *homotopy category* of $\mathbf{C}$.

Assume now that $*$ is an initial object of $\mathbf{C}$ and let $\mathbf{C}_{cf}$ be the full subcategory of $\mathbf{C}$ which consists of cofibrant and fibrant objects in $\mathbf{C}$; a morphism in $\mathbf{C}_{cf}(A, U)$ is an element of $[A, U]$. Note that by definition 5.1.1 a morphism $X \longrightarrow Y$ in $\mathrm{Ho}(\mathbf{C})$ can be represented by a chain of morphisms in $\mathbf{C}$:

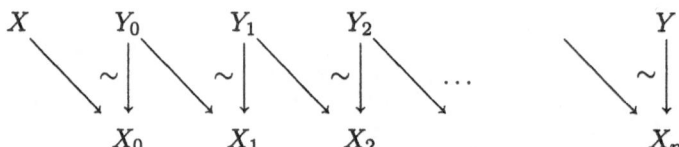

where the vertical arrows are weak equivalences in $\mathbf{C}$ (the name is justified in §5.2).

5.1.2 Remark. Let $\mathbf{Z}$ and $\mathbf{K}$ be categories with initial objects and let

$$\alpha\colon \mathbf{Z} \longrightarrow \mathbf{K}$$

be a functor which carries weak equivalences to weak equivalences. The universal property of $\mathrm{Ho}(\mathbf{Z})$ implies that α induces a functor

$$\mathrm{Ho}(\alpha)\colon \mathrm{Ho}(\mathbf{Z}) \longrightarrow \mathrm{Ho}(\mathbf{K})$$

between homotopy categories.

Then we get the following.

5.2 Theorem. Homotopy relative to $*$ is a natural equivalence relation on $\mathbf{C}_{cf}$ and there is an equivalence of homotopy categories

$$RM\colon \qquad \mathrm{Ho}(\mathbf{C}) \overset{\sim}{\dashleftarrow} \mathbf{C}_{cf}/(\simeq \mathrm{rel}\,*).$$

Proof. First we construct the functor

$$RM\colon \mathbf{C} \longrightarrow \mathbf{C}_{cf}/(\simeq \mathrm{rel}\,*).$$

For each object X in $\mathbf{C}$ we choose models RX, MX and RMX, where RX (resp. MX) denotes the fibrant (resp. cofibrant) model of X. Then we obtain for $f\colon X \longrightarrow Y$ in $\mathbf{C}$ the following diagram:

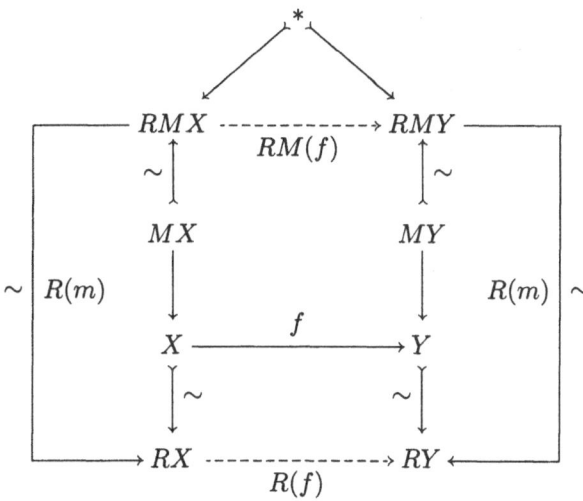

In this diagram the existence of $R(f)$ and $R(m)$ is ensured by the extension lemma 3.1, and $RM(f)$ lifts $R(f)$ as in the lifting lemma 3.3 (consider the outer arrows).

The homotopy class of $RM(f)$ is well defined by f. Let $H(\mathbf{C})$ be the category with the same objects as $\mathbf{C}$ and whose morphisms are defined by

$$H(\mathbf{C})(X,Y) \underset{\text{Def}}{=\!=\!=} [RMX, RMY]$$

For $f\colon X \longrightarrow Y$ let $q(f)$ denote the homotopy class of $RM(f)$ rel $*$. Then the functor $q\colon \mathbf{C} \longrightarrow H(\mathbf{C})$ has the universal property stated in definition 5.1.1 and therefore $H(\mathbf{C}) = \mathrm{Ho}(\mathbf{C})$.

Moreover the diagram above gives a functor $\mathbf{C} \longrightarrow \mathbf{C}_{cf}/(\simeq \mathrm{rel}\,*)$ which induces the inverse of the functor $RM\colon \mathrm{Ho}(\mathbf{C}_{cf}) \xrightarrow{\mathrm{Ho}(i)} \mathrm{Ho}(\mathbf{C})$ (where i is the inclusion $\mathbf{C}_{cf} \subset \mathbf{C}$). $\qquad\qquad\qquad\qquad\qquad\qquad\qquad\qquad\qquad\quad$ $\square$

Chapter 4
Algebraic Examples of Cofibration Categories

Up to now we have mentioned only one example of a cofibration category, namely topological spaces. Actually, the notion of cofibration category is an attempt to axiomatize the minimal properties to get a "good" homotopy theory. This chapter gives some algebraic instances of cofibration categories. We focus our attention on two cases, the categories **CDA** and **DL**, which are of particular importance because of their connections with topology. We resume all the relevant definitions in Table 1.

Preliminary remark: Let k be a field (often specialized to $\mathbb{Q}$) and consider now k-vector spaces. All graded vector spaces will be written $V = V^*$, with $V^i = 0$ if $i < 0$. If V is a graded vector space we write $V^{<k} = \{V^i\}_{i<k}$. We shall often consider a basis $\{v_\alpha\}_{\alpha \in I}$ of V indexed by a well ordered set I, then $V_{<\alpha}$ will denote the span $\{v_\beta\}_{\beta<\alpha}$.

Moreover, if V is a k-vector space we denote by $\underline{V}$ the graded vector space defined by $\underline{V}_i = k$ if $i = 0$ and $\underline{V}_i = 0$ if $i \neq 0$.

Table 1: Algebraic examples of cofibration categories

Notation	$\textbf{Chain}_R^+$	**DA**	**DL**	$\textbf{CDA}_*^0$
Ring	R any ring	R principal domain	R field of char. 0	R field of char. 0
Objects	R-chain complexes	chain algebras over R, flat (or free) as R-module	chain Lie algebras over R	comm. cochain algebras over R, connected augmented
Morphisms	chain maps	chain algebra maps	chain Lie algebra maps	cochain algebra maps, augmented
we	homology isomorphisms			
cof	injection with proj. cokernel	free extension $X \to X \amalg T(V)$	free extension $X \to X \amalg \mathbb{L}(V)$	KS-extension $X \to X \otimes \Lambda(V)$
fibrant	all objects			

§1   The category CDA

1.1 Definition. A *commutative differential algebra* on $\mathbb{Q}$ is a graded $\mathbb{Q}$-vector space $A = (A^n)_{n \in \mathbb{N}}$ equipped with an associative multiplication m with unit $1 \in A^0$ and a derivation d (i.e. $d \circ m = m \circ (d \otimes A + A \otimes d)$) of degree $+1$ which is also a *differential* ($d \circ d = 0$) and such that the following diagram commutes:

$$
\begin{array}{ccc}
A \otimes A & \xrightarrow{\ m\ } & A \\
{\scriptstyle T}\Big\downarrow & \nearrow{\scriptstyle m} & \\
A \otimes A & &
\end{array}
$$

where $T(a \otimes b) = (-1)^{|a| \cdot |b|} (b \otimes a)$ for all a, b.

We call **CDA** the category of commutative differential algebras (morphisms are the obvious ones).

The free CDA generated by A is denoted by $\Lambda(A)$ and defined as the tensor product of the polynomial algebra on A^{even} and the exterior algebra on A^{odd}. We use the notation $\Lambda^{<p}V = \underset{k<p}{\oplus} \Lambda^k V$.

Remark. $\mathbb{Q}$ has a trivial structure of CDA and is an initial object in the category.

1.2 Sullivan models.

1.2.1 Definition. A *Sullivan model* is a CDA of the form $(\Lambda V, d)$ such that V has a well ordered basis $\{v_\alpha\}$, $\alpha \in J$, with the property that

$$(*) \qquad\qquad dv_\alpha \in \Lambda V_{<\alpha} \quad \text{and} \quad H^0(\Lambda V) = \mathbb{Q}.$$

A special case of importance is when V has a basis $v_1, v_2, \ldots$ and $dv_i \in \Lambda(v_1, \ldots, v_{i-1})$, $dv_1 = 0$ (of course dv_i is a cycle).

Remark. The inclusion

$$\Lambda(v_1, \ldots, v_{i-1}) \rightarrowtail \Lambda(v_1, \ldots, v_{i-1}) \otimes \Lambda(v_i)$$

is called an *elementary cofibration* in **CDA**. More generally, let $A \in ob(\textbf{CDA})$, and $z \in ZA_n$ a cycle. We define the algebra $A \otimes \Lambda(a)$, with a of degree $n+1$ and a differential D such that $Da = z$; then the morphism $A \rightarrowtail A \otimes \Lambda(a)$ is called an elementary cofibration in **CDA**.

Observe that $\text{Im}\, d \subset (\Lambda V)^{\geq 1} \subset \Lambda^{\geq 1}V$. If actually $\text{Im}\, d \subset \Lambda^{\geq 2}$ then the model is called *minimal model*. Because d is a derivation this is equivalent to: $d(V) \subset \Lambda^{\geq 2}V$.

1.2.2 Lemma. Let $(\Lambda V, d)$ be a minimal model. Then

i) $V = V^{\geq 1}$.

ii) $d: V^k \longrightarrow \Lambda V^{\geq k}$.

iii) $V^i = 0$, for $i = 1, \ldots, r$ iff $H^i(\Lambda V) = 0$, for $i = 1, \ldots, r$.

iv)   If $V^1 = 0$ and $V^i = 0$ for $i = 1, \ldots, r-1$ then $V^r = H^r(\Lambda V)$.

v)   If $V^{\geq 1} = 0$, $d: V^k \longrightarrow \Lambda V^{<k}$.

vi)   Suppose $V^1 = 0$, then V (and ΛV) have finite type iff $H(\Lambda V)$ does.

Proof. i) If $V^0 \neq 0$, let v_i be the minimal vector basis of degree zero in the expression

$$dv_i = \Sigma \lambda_{jk} v_j \wedge v_k + \Sigma \lambda_{jkl} v_j \wedge v_k \wedge v_l + \ldots$$

All v_j, v_k have degree ≥ 1, and the right hand side has degree ≥ 2, therefore it is zero and since $H^0 = \mathbb{Q}$, the element $\psi 1 + \sigma v_i$ must be the boundary of an element of degree -1, which is impossible.

ii)–v) Can be proved along the same lines and are left to the reader as an exercise.

vi) Clearly, ΛV has finite type if V does, and $H(\Lambda V)$ has finite type if ΛV does. Conversely, assume $\dim V^i < \infty$ if $i < r$ and $\dim V^r = \infty$. Since $\operatorname{Im} d \subset \Lambda^{\geq 2} V$ one has $d(V^r) \subset [\Lambda(V^{<r})]^{r+1}$ which is finite dimensional. Therefore $K^r = \operatorname{Ker} d/V^r$ is infinite dimensional and contains no non-zero boundary. Thus K^r injects in H^r and the latter is infinite dimensional.   □

The most important special case of condition $(*)$ in Definition 1.2.1 is when V has a basis of the form

$$v_1, v_2, \ldots, v_r \quad \text{with} \quad dv_r \in \Lambda(v_1, \ldots, v_{r-1}).$$

Note that necessarily the first vector verifies $dv_1 = 0$. We make the following observations in this case:

i)   Because d is a derivation, the Sullivan model is entirely determined by the elements

$$z_{r+1} = dv_{r+1} \in \Lambda(v_1, \ldots, v_r).$$

Moreover, d maps each $\Lambda(v_1, \ldots, v_r)$ into itself, i.e. each $(\Lambda(v_1, \ldots, v_r), d)$ is itself a Sullivan model and z_{r+1} is a cycle (not necessarily a boundary) in $\Lambda(v_1, \ldots, v_r)$.

ii) Conversely, suppose we specify arbitrary elements z_{r+1} in $\Lambda(v_1, \ldots, v_r)$ such that $dv_{r+1} = z_{r+1} + 1$. Then there is a unique derivation δ of degree 1 in ΛV such that

$$\delta V_1 = 0 \quad \text{and} \quad \delta v_{r+1} = z_{r+1}.$$

Moreover, $\delta^2 = \frac{1}{2}(\delta\delta - (-1)^{|\delta||\delta|}\delta\delta) = \frac{1}{2}[\delta, \delta]$ is again a derivation and so $\delta^2 = 0$ if and only if $\delta z_{r+1} = 0$. Since $z_{r+1} \in \Lambda(v_1, \ldots, v_r)$ we can compute δz_{r+1} from the elements $\delta v_i = z_i$, $1 \leq i \leq r$. Thus we have an inductive way of checking that $\delta^2 = 0$.

These observations give an easy way to construct Sullivan models. We start with $(\Lambda v_1, d)$ with $dv_1 = 0$. Assume $(\Lambda(v_1, \ldots, v_r), d)$ has been built; we add a new variable v_{r+1} with $\deg v_{r+1} = \deg z_{r+1} - 1$ and set $dv_{r+1} = z_{r+1}$. Repeat the process, possibly at infinitum.

N.B. We are always allowed to choose $z_{r+1} = 0$. This just means $dv_{r+1} = 0$; i.e. we have added a new cycle to the model.

1.2.3 Examples.
i) $\Lambda(x, y, z)$ with $dx = dy = dz = 0$, $\deg x = \deg y = \deg z = 3$. Then xyz is a cycle of degree 9.
ii)   Introduce a new variable, a, of degree 8 with $da = xyz$.
      Consider $\Lambda(x, y, z, a)$, then xya is a cycle of degree 14.
iii) Introduce u with $\deg u = 13$ and $du = xya$.
      Consider $\Lambda(x, y, z, a, u)$, then $a^2 + 2zu$ is a cycle of degree 16.
iv)   Introduce v with $\deg u = 15$ and $dv = (a^2 + 2zu)$.
      Consider $(\Lambda(x, y, z, a, u, v), d)$, then we shall prove in chapter 7, §3 that $H^i(\Lambda(x; y, z,$
      $a, u, v)) = 0$ if $i > 30$ and $\dim H^{30} = 1$.

1.3 Cofibrations in CDA.
Sullivan models $(\Lambda V, d)$ as defined in the above paragraph are endowed with a natural map $\varepsilon: (\Lambda V, d) \longrightarrow \mathbb{Q}$ defined by $\varepsilon(V) = 0$. We say that a CDA A is augmented when a CDA map $\varepsilon: A \longrightarrow \mathbb{Q}$ is defined.

1.3.1 Definition. Let $\mathbf{CDA}_*$ be the category of CDAs over $\mathbb{Q}$ (an object in $\mathbf{CDA}_*$ is a map $\varepsilon: A \longrightarrow \mathbb{Q}$ called an augmentation).

1.3.2 Definition. A map $B \longrightarrow A$ in $\mathbf{CDA}_*$ is a cofibration (or *KS-extension*) if there is a subspace V of A and a well-ordered subset J_V of V such that
i) V is a vector space with basis J_V and if ε is the augmentation, $\varepsilon(V) = 0$.
ii)   The homomorphism $B \otimes \Lambda(V) \longrightarrow A$ of commutative algebras, given by $B \longrightarrow A$ and $V \subset A$, is an isomorphism of algebras.
iii) For $\alpha \in J_V$, write $V_{<\alpha}$ for the subspace of V generated by all $\beta \in J_V$, with $\beta < \alpha$. Then the differential in A satisfies $d(\alpha) \in B \otimes \Lambda(V_{<\alpha})$ where we use the isomorphism in ii).

Remark: A cofibration $\mathbb{Q} \longrightarrow A$ is precisely what we called a Sullivan model in 1.2.1.

1.4 Acyclicity.

1.4.1 Definition. A map $f: A \longrightarrow A'$ in $\mathbf{CDA}$ is a *weak equivalence* if and only if its homology $H(f): H(A) \longrightarrow H(A')$ is an isomorphism (we say also that f is quasi-isomorphism).

In differential categories acyclic objects play an important role. We recall that a graded differential object A in a category with an initial object $*$ is *acyclic* if and only if its homology is concentrated in degree zero and equal to $H(*)$ (that is A is quasi-isomorphic to $*$). The next proposition characterizes acyclicity in CDA.

1.4.2 Proposition. Let V be a graded $\mathbb{Q}$-vector space. The CDA $(\Lambda(V \oplus dV), d)$ where dV is defined by $(dV)^{n+1} = V^n$ and d by the shift of degree on V and $d(dV) = 0$ is acyclic. The proposition holds for CDAs over a field k if and only if the characteristic of k is zero.

Proof. Consider $\Lambda(V \oplus dV)$ with the differential defined above; let s be the shift of degree -1

$$s: dV \longrightarrow V$$

$$s: V \longrightarrow 0.$$

Extend s to a derivation of degree -1 and consider the derivation of degree 0 defined by $\Delta = [s, d] = sd + ds$. Δ is the identity on $V \oplus dV$, and therefore the multiplication by the length of words on $\Lambda(V \oplus dV)$.

If char $k = 0$, Δ is an isomorphism and the proposition is proved. □

1.5 Push-outs. We first point out the following particular case. If A and B are objects in CDA, with respective multiplications m_A and m_B then the following diagram defines a multiplication on the tensor product $A \otimes B$.

$$
\begin{array}{ccc}
A \otimes B \otimes A \otimes B & \xrightarrow{\quad m \quad} & A \otimes B \\
T \downarrow & \nearrow m_A \otimes m_B & \\
A \otimes A \otimes B \otimes B & &
\end{array}
$$

The algebra $A \otimes B$ is the coproduct of A and B in the category **CDA**.

Let us recall that $m: A \otimes A \longrightarrow A$ is a map of algebras if and only if A is commutative.

1.5.1 Definition. Let $i: (B \longrightarrow (B \otimes \Lambda(V), d))$ be a cofibration in **CDA**$_*$ and $f: B \longrightarrow Y$ a map. Then we define a cocartesian diagram:

$$
\begin{array}{ccc}
B & \xrightarrow{\quad f \quad} & Y \\
i \downarrow & & \downarrow i \\
B \otimes (\Lambda(V), d) = A & \xrightarrow[\quad \bar{f} \quad]{} & A \cup_B Y = (Y \otimes \Lambda(V), d)
\end{array}
$$

by $\bar{f} = f \otimes T(V)$. The differential d on the push-out $A \cup_B Y = Y \otimes (\Lambda V, d)$ is defined by the condition that f and i are differential morphisms.

1.6 Cofibration category. The category **CDA**$_*$ might appear to be a good candidate to be a cofibration category; nevertheless the factorization axiom (Axiom C_3) fails. So we are led to consider the category **CDA**$^0_*$ defined as the full subcategory of **CDA**$_*$ consisting of all *connected* augmented CDAs, that is augmented CDAs $\varepsilon: A \longrightarrow \mathbb{Q}$ where $H^0\varepsilon: H^0 A \longrightarrow H^2\mathbb{Q}$ is an isomorphism.

1.6.1 Proposition. **CDA**$^0_*$ is a cofibration category.

Proof. Weak equivalences and cofibrations are defined as for $\mathbf{CDA}_*$. The definition of push-outs is more complicated, but then the factorization axiom holds. For a detailed proof see [8], Chapter 1.   □

In the category $\mathbf{CDA}^0_*$ (and in $\mathbf{CDA}_*$ also) all objects are fibrant.

1.6.2 Definition. Let $f\colon A \longrightarrow B$ be a morphism in $\mathbf{CDA}^0_*$. Then every factorization (Axiom C3) of f is called a *(cofibrant) model* of f. If $*$ is the initial object of $\mathbf{CDA}^0_*$, the factorization of $* \longrightarrow B$ is called a ($*$-cofibrant) model of B.

So a cofibrant model is a free extension. In particular the $*$-cofibrant objects are the free objects.

1.7 Cylinders. We intend to characterize homotopy in our category $\mathbf{CDA}_*$. For this purpose we need an explicit construction of the cylinder.

1.7.1.   For simplicity we restrict ourselves to Sullivan models. So consider the Sullivan model $\mathbb{Q} \longrightarrow \Lambda V$.

Let us now define the Sullivan model (duplication of ΛV): $\Lambda V \otimes_\mathbb{Q} \Lambda V (\cong \Lambda(V_1 \oplus V_2)$, with $V_1 \cong V \cong V_2$) equipped with two evident cofibrations

$$\Lambda V \xrightarrow[\;i_2\;]{\;i_1\;} \Lambda V \otimes_\mathbb{Q} \Lambda V;$$

we define the folding map

$$\pi\colon \Lambda V \otimes_\mathbb{Q} \Lambda V \cong \Lambda(V_1 \oplus V_2) \longrightarrow \Lambda V$$

where $\pi/V_i\colon V_i \longrightarrow V$, $i = 1, 2$, are the isomorphisms above.
We proceed to construct an explicit factorization of π:

$$\Lambda V \otimes_\mathbb{Q} \Lambda V \xrightarrow{\;i\;} I_\mathbb{Q}\Lambda V \xrightarrow{\;p\;} \Lambda V$$

where $H(p)\colon H^*(I_\mathbb{Q}\Lambda V) \longrightarrow H^*(\Lambda V)$ is an isomorphism (we say that $H(p)$ is a *weak equivalence*).

Let $V^+ = V/V^0$ and let sV^+ be the graded vector space defined by $(sV^+)^{n-1} = (V^+)^n$, and denote also by s the map which is the composition of the projection $V \longrightarrow V/V^0$ followed by the isomorphism of degree -1: $(V^+)^n = (sV^+)^{n-1}$.

We define the cylinder over ΛV as

$$I_\mathbb{Q}\Lambda V = \Lambda V \otimes \Lambda(sV^+ \oplus dsV^+),$$

where $\Lambda(sV^+ \oplus dsV^+)$ is acyclic (cf. §1.4). Furthermore define

$$p = \Lambda V \otimes \varepsilon$$

and $j = \Lambda V \otimes \Lambda V \longrightarrow I_\mathbb{Q}\Lambda V$ by $j = j_0 \otimes j_1$ with $j_0 = \Lambda V \otimes 1$.
The definition of j_1 is more complicated.

First a derivation S of degree -1 is defined on $I_Q \Lambda V$ by

$$S(sV^+) = 0 = S(dsV^+)$$
$$S(v) = \bar{s}(v) \in sV^+ \quad \text{for} \quad v \in V.$$

Then $\theta = dS + Sd$ is a degree zero derivation. One checks that for every x in $I_Q \Lambda V$ there is an integer N such that $\theta^N(x) = 0$. Then the expression $e^\theta = \sum_0^\infty \frac{\theta^n}{n!}$ is an endomorphism of $I_Q \Lambda V$. Obviously it is an automorphism and one defines $j_1 = e^\theta j_0$.

We are now ready to define the notion of homotopy.

1.7.2 Homotopy. Let $(\Lambda V, d)$ be a Sullivan model and A a CDA. We first want to briefly look at two cases of particular importance.

Suppose $A = \Lambda v$ with $|v|$ odd and differential $d = 0$. Let $f, g \colon \Lambda V \longrightarrow A$ be two homotopic maps in $\mathbf{CDA_*}$, i.e. there is H such that the diagram

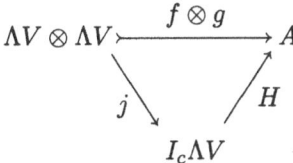

commutes.

Choosing a basis $\{v_\alpha\}_\alpha$ of V, the maps f (resp. g) are determined by the rationals $f(\alpha)$ (resp. $g(\alpha)$) with $f(v_\alpha)f(\alpha)v$ (resp. $g(v_\alpha) = g(\alpha)v$) where $f(\alpha)$ (resp. $g(\alpha)$) are zero if $|v_\alpha| \neq |v|$. An easy calculation proves that the existence of the factorization H is equivalent to $f(\alpha) = g(\alpha)$. In conclusion we can identify the "homotopy classes" of applications $\Lambda V \longrightarrow \Lambda v$, $|v| = 2n+1$ with $\mathrm{Hom}(V_{2n+1}, \mathbb{Q})$.

The case $A = \Lambda(v, w)$, $|v| = 2n$, $|w| = 4n - 1$ and $dv = 0$, $dw = v^2$, though less simple to handle with leads to the analogue conclusion.

We will see in chapter 5 that the two types of Sullivan models considered above are algebraic models of respectively odd and even dimensional rational spheres. This justifies the coming definitions.

1.8 Homotopy groups of the minimal model. Let $(\Lambda V, d)$ be a minimal model. It is important to observe that there is no canonical choice for the generating subspace V.

Example. Let $\Lambda(a, x, y, u, v)$ with $da = dx = dy = 0$, $du = a^2$, $dv = xy - a^3$.
Apply the following change of basis:

$$a' = a, \ x' = x, \ y' = yu' = u, \ v' = v + au,$$

then $\Lambda(a, x, y, u, v) = \Lambda(a', x', y', u', v')$, with $da' = dx' = dy' = 0$, $du' = (a')^2$, $dv' = x'y'$. Thus there is no canonical basis nor differential. But we have:

1.8.1 Proposition. The filtration of $(\Lambda V, d)$ defined by:

$$(\Lambda V)^+ = \Lambda^{\geq} V, \ (\Lambda V)^+ (\Lambda V)^+ = \Lambda^{\geq 2} V, \ \text{etc} \ldots$$

is functorial.

Remark. $\Lambda^1 V / \Lambda^{\geq 2} V$ is isomorphic to V. In fact the graded module associated to this filtration is (non canonically) isomorphic to $(\Lambda V, d)$. The corresponding graduation is called the words length graduation and we write after a choice of V:

$$\Lambda V \cong \mathbb{Q} \oplus V \oplus \Lambda^2 V \oplus \Lambda^3 V \oplus \ldots$$

Modulo this decomposition the differential can be written as

$$d = d_1 + \gamma \quad \text{with} \quad d_1 \colon \Lambda^k V \longrightarrow \Lambda^{k+1} V$$

$$\gamma \colon \Lambda^k V \longrightarrow \Lambda^{\geq k+2} V.$$

where
i) d_1 and γ are derivations (γ is not functorial).
ii) d_1 is a differential: $(d_1)^2 = 0$.   □

1.8.2 Definition. Let $\pi^*(\Lambda V, d) = \Lambda^{\geq 1} V / \Lambda^{\geq 2} V$. Then we define $\pi_*(\Lambda V, d)$, the *homotopy* of the Sullivan minimal model $(\Lambda V, d)$ by

$$\pi_*(\Lambda V, d) = \operatorname{Hom}(\pi^*(\Lambda V, d), \mathbb{Q}) \ (\cong \operatorname{Hom}(V, \mathbb{Q})).$$

We may think of $\pi_*(\Lambda V, d)$ as the linear functions on $(\Lambda V)^2$ that vanish on $\Lambda^{\geq 2}$.
    Note that $\pi_*(\Lambda V, d)$ is a contravariant functor of $(\Lambda V, d)$.

1.8.3 Whitehead product. Definition. Let $(\Lambda V, d)$ be a minimal Sullivan model. $\pi_*(\Lambda V)$ has a *Whitehead product*:

$$[\,;\,] \colon \pi_k \times \pi_l \longrightarrow \pi_{k+l-1}.$$

defined as follows:
    Let f and g be maps $(\Lambda^{\geq 1} V, \Lambda^{\geq 2} V) \longrightarrow (\mathbb{Q}, 0)$ of respective degree k and l. We want to construct a map

$$[f, g] \colon (\Lambda^{\geq 1} V, \Lambda^{\geq 2} V) \longrightarrow (\mathbb{Q}, 0)$$

of degree $k + l - 1$.
    Let $f \otimes g \colon (\Lambda^{\geq 2}, \Lambda^{\geq 3}) \longrightarrow (\mathbb{Q}, 0)$ be defined by

$$f \otimes g(\Phi \wedge \Phi) = g(\Phi)f(\Psi) + (-1)^{|g| \cdot |\Phi|} \ g(\Psi)f(\Phi).$$

Then we define:

$$\langle [f, g], \Phi \rangle = \langle f \otimes g, d\Phi \rangle.$$

1.8.4 Remark. Let $x \in V$ with $dx = \Sigma x_i \wedge y_i \bmod \Lambda^{\geq 3}V$. We have:

$$\langle [f, g], x \rangle = \langle f \otimes g, dx \rangle = \Sigma \langle f \otimes g, x_i \wedge y_i \rangle$$
$$= \pm \Sigma \langle f, x_i \rangle \langle g, y_i \rangle \pm \Sigma \langle f, y_i \rangle \langle g, x_i \rangle.$$

Thus $[f, g]$ is completely determined by d_1.

1.8.5 Remark. Define $L(\Lambda V, d)$ as the desuspension of $\pi_*(\Lambda V)$ $(L(\Lambda V, d)_n = \pi_{n+1}(\Lambda V))$. Then the Whitehead product provides $L(\Lambda V, d)$ with a structure of a graded Lie algebra.

Moreover, suppose now that $\varphi : (\Lambda V, d) \longrightarrow (\Lambda W, d)$ is a morphism of **CDA**. Then, by restriction, we can define the linear part

$$\varphi_1 : \Lambda^1 V \longrightarrow \Lambda^1 W.$$

and φ_1 commutes with $d_1 : \varphi_1 d_1 = d_1 \varphi_1$. Call $L(\varphi)$ the induced map

$$L(\varphi) : L_{(\Lambda V, d)} \longleftarrow L_{(\Lambda W, d)}.$$

Now the definition of the Whitehead product shows that $L(\varphi)$ is Lie algebra homomorphism.

1.8.6 Lemma. φ is surjective if and only if $L(\varphi)$ is injective. □

1.8.7 Examples. In the previous example 1.2.3, namely

$$\Lambda(x, y, z, a, u, v), \ da = xyz, \ du = xya, \ dv = a^2 + 2zu,$$

π_* has a basis $x^*, y^*, z^*, a^*, u^*, v^*$ of respective degrees 3, 3, 3, 8, 13, 15. The quadratic part of the differential is given by:

$$d_1 x = 0, \ d_1 y = 0, \ d_1 z = 0, \ d_1 a = 0, \ d_1 u = 0, \ d_1 v = a^2 + 2zu.$$

Therefore all Whitehead product are zero except:

$$[z^*, u^*] = \pm 2v^*$$
$$[a^*, a^*] = \pm 2v^*.$$

We now describe very briefly the categories Chain$_R^+$ and **DA**. We refer to Table 1 at the beginning of this chapter for a list of all definitions.

§2 The category Chain$_R^+$

2.1 Definition. Let R be a unitary ring. A *chain complex* V is a graded R-module $(V_n)_{n \in \mathbb{Z}}$ equipped with a differential $(d \circ d = 0)$ of degree -1.

We call **Chain$_R$** the category whose objects are chain complexes (of course morphisms should be compatible with the differential).

2.2 Weak equivalences, cofibrations and push-outs.

2.2.1 Definitions. A map $f: V \longrightarrow V'$ in $\mathbf{Chain}_R$ is a *weak equivalence* if and only if its homology $H(f): H(V) \longrightarrow H(V')$ is an isomorphism (we also say then that f is a *quasi- isomorphism*).

A map $i: V \longrightarrow V'$ in $\mathbf{Chain}_R$ is a cofibration if and only if there is a free submodule W of V' and an isomorphism of graded modules $V' \cong i(V) \oplus W$.

Let $i: V \longrightarrow V \oplus W$ be a cofibration and $f: V \longrightarrow U$ any map in $\mathbf{Chain}_R$. Then we define a cocartesian diagram

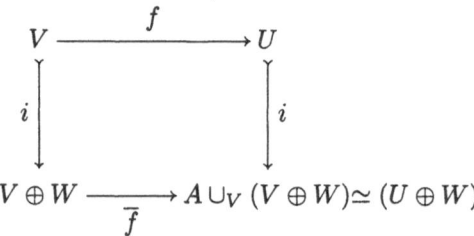

by $\bar{f} = f \oplus W$. The differential on $U \cup_V (V \oplus W)$ is defined by requiring $\bar{f}$ and $\bar{i}$ to be differential morphisms.

2.3 Let $\mathbf{Chain}_R^+$ be the following subcategory of $\mathbf{Chain}_R$: a chain complex V is an object of $\mathbf{Chain}_R^+$ if and only if there is an $N \in \mathbb{Z}$ such that $V_n = 0$ for $n < N$.

Proposition. The category $\mathbf{Chain}_R^+$ is a cofibration category in which all objects are fibrant.

Proof. Weak equivalences, cofibrations and push-outs are defined as in 2.3 for $\mathbf{Chain}_R$. It essentially remains to check the factorization axiom. We leave this to the reader. $\square$

2.4 Cofibrant models. We first observe that the trivial chain complex is both an initial and a final object in $\mathbf{Chain}_R^+$.

Let X be a (right) R-module and $M_* \overset{\sim}{\longrightarrow} X$ be a R-free resolution of X. Then $0 \rightarrowtail M_* \overset{\sim}{\longrightarrow} X$ is a 0-cofibrant model of X in $\mathbf{Chain}_R^+$. Recall that Tor and Ext are defined by

$$\mathrm{Tor}^{R_i}(X, Q) = H_i(MX \otimes_Q N)$$

$$\mathrm{Ext}_{R_i}(X, P)) = H^i(\mathrm{Hom}(MX, P))$$

where Q (resp. P) is a left (resp. right) R-module. The uniqueness of MX shows that Tor and Ext are well defined. The lifting lemma (3.3, chapter 3) shows that Tor and Ext are functors.

More generally, if X and Y are complexes in $\mathbf{Chain}_R^+$ this gives a definition of the differential Tor and Ext.

Remark. All of what we said about $\mathbf{Chain}_R$ and $\mathbf{Chain}_R^+$ remains valid if we define a cofibration $I: V \longrightarrow V'$ by the existence of a projective R-module W with $V' \cong i(V) \oplus W$.

§3 The category DA

3.1 Definition. Let k be a field. The category of *differential algebras* (denoted by **DA**) has as objects the graded vector spaces $A_* = (A_n)$, $n \in \mathbb{N}$, equipped with an associative multiplication $m: A \otimes A \longrightarrow A$ of degree 0, with a unit $1 \in A_0$ ($m(1 \otimes x) = m(x \otimes 1) = x$), and with a derivation (i.e. $d \circ m = m \circ (d \otimes A + A \otimes d)$) of degree -1 which is also a differential. The morphisms are the obvious ones.

Let k be the DA defined by $k_i = k$ if $i = 0$, and $k_i = 0$ if $i \neq 0$. There is a unique map of unitary algebras $k \longrightarrow A$ and k is an initial object; A is *augmented* if there is a retraction $\varepsilon: A \longrightarrow k$. If V is a k-vector space, one defines the free algebra on V by the tensor algebra $T(V)$.

Examples.
i) For G a topological group, the singular chain complex $C_*(G; k)$ is an object in **DA**.
ii) The *Moore loop space* ΩX of a pointed space $(X, *)$ is defined by:

$$\Omega X = \{(\lambda, T) \in X^{\mathbb{R}^+} \times \mathbb{N}; \lambda: \mathbb{R}^+ \longrightarrow X \text{ with } \lambda(0) = \lambda(t) = *, \text{ for } t \geq T\}.$$

ΩX is a topological monoïd with composition law defined by

$$(\lambda, S) \circ (\mu, T) = (\lambda\mu, S + T) \quad \text{with} \quad \lambda\mu(s) = \begin{cases} \lambda(s), & \text{if } s \leq S \\ \mu(s - S), & \text{if } s \geq S; \end{cases}$$

so $C_*(\Omega X)$ is an associative algebra.

By the correspondance $X \longrightarrow C_*(\Omega X)$ we get a functor **Top** $\longrightarrow$**DA**.

3.2 Weak equivalences. Definition. A map $f: A \longrightarrow A'$ in **DA** is a *weak equivalence* if and only if its homology $H(f): H(A) \longrightarrow H(A')$ is an isomorphism (we say also that f is a quasi-isomorphism).

3.3 Cofibrations. Let A and B be augmented objects in **DA**. Let $\overline{A} = \ker \varepsilon$ be the augmentation ideal.

We define the *coproduct* of A and B:

$$A \amalg B = \bigoplus_{k \geq 0} A \otimes (\overline{B} \otimes \overline{A})^{\otimes k} \otimes B.$$

This object has a differential algebra structure in an evident way; by the Künneth theorem one easily sees that $HA \amalg HB = H(A \amalg B)$.

3.3.1 Definition. Let $A \in$ ob **DA**, and $z \in ZA_n$ (as usual the cycles of A_n). Let a be an element of degree $n + 1$, and let $T(a)$ be the free associative k-algebra generated by a; define on the coproduct $A \amalg T(a)$ a differential D such that $Da = z$. Then the morphism $A \subset A \amalg T(a)$ is called an *elementary cofibration* in **DA**.

This procedure is topologically significant and corresponds to attaching a cell to a space. It is the starting point of the theory of Adams-Hilton. Let us very briefly recall that here.

Attaching an $n + 2$-cell to a space X is defined by the following push-out square.

$$
\begin{array}{ccc}
\sum S^n = S^{n+1} & \xrightarrow{\ f\ } & X \\
\downarrow & & \downarrow{\scriptstyle j} \\
e^{n+2} & \longrightarrow & X \cup_f e^{n+2}
\end{array}
$$

By dualizing f, we define $\bar{f}\colon S^n \longrightarrow \Omega X$ and

$$H_*\bar{f}\colon H_*(S^n; \mathbb{Z}) \longrightarrow H_*(\Omega X; \mathbb{Z})$$

If i_n is a generator of $H_n(S^n; \mathbb{Z})$ choose $z_n \in C_n(\Omega X; \mathbb{Z})$ such that $\bar{z}_n = \bar{f}(i^n)$.

3.3.2 Theorem (Adams-Hilton 1956). Let us define the DA map φ, for $|a| = n + 1$:

$$\varphi\colon C_*(\Omega X) \amalg T(a) \longrightarrow C_*(\Omega(X \cup e^{n+2}))$$

by $da = z_n$, $\varphi = C_*(\Omega j)$ on $C_*(\Omega X)$ and $\varphi(a) = b$ with $db = C_*(\Omega j)(z_n)$. Then φ is a quasi-isomorphism.

Proof. See [1].   □

The process indicated in the theorem provides each CW-complex X with a differential algebra A_X, which is a cofibrant model (cf. 3.5.2) in **DA** of $C_*(\Omega X)$. A_X is called the *Adams-Hilton model* of X.

3.3.3 Definition. A map $i\colon B \longrightarrow A$ in **DA** is a *cofibration* if and only if there is a free subspace V of A and an isomorphism of algebras $A \cong i(B) \oplus T(V)$.

Let $i\colon B \longrightarrow (B \amalg T(V), d)$ be a cofibration and $f\colon B \longrightarrow C$ any map in **DA**. Then we define a cocartesian diagram

$$
\begin{array}{ccc}
B & \xrightarrow{\ f\ } & Y \\
{\scriptstyle i}\downarrow & & \downarrow{\scriptstyle \bar{i}} \\
B \amalg (T(V), d) = A & \xrightarrow[\ \bar{f}\ ]{} & A \cup_B Y (Y \amalg T(V), d)
\end{array}
$$

by $\bar{f} = f \amalg T(V)$. The differential on $C \amalg T(V)$ is defined by requiring $\bar{f}$ and $\bar{i}$ to be differential morphisms.

3.4 Acyclicity. Proposition. The differential algebra $(T(V \oplus dV), d)$, where dV is defined by $(dV)_n = V_{n+1}$ and the differential $d\colon V \longrightarrow dV$ is defined by the shift of degree -1 on V and $d(dV) = 0$, is acyclic.

Proof. 1°) Consider $T(V \oplus dV)$ with the differential defined above; let s be the shift of degree $+1$

$$s: dV \longrightarrow V$$
$$s: V \longrightarrow 0.$$

Extend s to a derivation of degree $+1$ and consider the derivation of degree 0 defined by $\Delta = [s, d] = sd + ds$. Δ is the identity on $V \oplus dV$, and therefore the multiplication by the length of words on T.

If char $k = 0$, Δ is an isomorphism and the proposition is proved.

2°) Proof for $(T(V \oplus dV), d)$, char k arbitrary.

Suppose now that V is an R-free module for R any ring. Decompose

$$T = T(V \oplus dV) = k \oplus V \otimes T \oplus dV \otimes T$$

and define S on T by

$$S(1) = 0, \quad S(v \otimes \Phi) = 0, \quad S(dV \otimes \Phi) = v \otimes \Phi.$$

Then

$$Sd + dS = id_T - \varepsilon,$$

where $\varepsilon: \Lambda(t, dt) \longrightarrow \Lambda(t, dt)$ is defined by $\varepsilon(t) = \varepsilon(dt) = 0$ and $\varepsilon(1) = 1$. That gives the result.   □

3.5 Cofibration category.

3.5.1 Theorem. The category **DA** is a cofibration category in which all objects are fibrant.

Proof. Cf. [8], chapter 1.   □

3.5.2 Cofibrant models. If we consider $\mathbf{DA}_*$ the full subcategory of augmented DAs (it is also a cofibration category), then $*$-cofibrant objects are the free objects.

§4   The category DL

4.1   Definition. A *differential Lie algebra* on $\mathbb{Q}$ is a graded vector space $L = (L_n)_{n \in \mathbb{N}}$ equipped with a bracket $[\,,]: L \otimes L \longrightarrow L$ and a derivation d (i.e. $d \circ [\,,] = [\,,] \circ (d \otimes A + A \otimes d)$) of degree -1 which is also a differential; $[\,,]$ has to satisfy the graded antisymmetric and Jacobi identities:

$$[a, b] = (-1)^{|a| \cdot |b|} [b, a]$$

$$[[a, b], c] = [a, [b, c]] + (-1)^{|a| \cdot |b|} [b, [a, c]] \text{ for all } a, b, c \text{ in } L.$$

We denote the corresponding category by **DL** (morphisms are the obvious ones).

If V is $\mathbb{Q}$-vector space, the free Lie Algebra $\mathbb{L}(V)$ is the span of all brackets of elements of V, subject to the antisymmetric and Jacobi identities.

Remark. On every differential algebra A, there is a canonical structure of (differential) Lie algebra; the bracket is defined by the formula:

$$[a, b] = m(a \otimes b) - (-1)^{|a| \cdot |b|} m(b \otimes a)$$

4.2 Weak equivalences, cofibrations and push-outs.

4.2.1 Definitions. A map $f: L \longrightarrow L'$ in **DL** is a *weak equivalence* if and only if its homology $H(f): H(L) \longrightarrow H(L')$ is an isomorphism (we say also that f is a quasi-isomorphism).

We define the *coproduct* in **DL** by using the definition for **DA**. First note that the *universal enveloping algebra* of a Lie algebra L is an algebra UL solving the following universal problem:

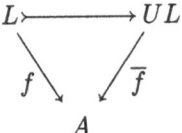

Let $A \in \mathrm{ob}\,\mathbf{DA}$, $f: \longrightarrow A$ a morphism in **DL** (cf. remark of §4.1); then $L \rightarrowtail UL$ is the injection in **DL** such that, for every f, there is a unique $\bar{f} \in \mathrm{Mor}_{\mathbf{DA}}(UL, A)$ making the diagram commute. (We can define $UL = T(L)/J$ where J is the ideal generated by all elements

$$[x, y] - (xy - (-1)^{|x| \cdot |y|} yx)).$$

Now we can state the important result:

Let L and $M \in \mathrm{ob}\,L$. One checks that the Lie algebra generated by L and M in the algebra $UL \amalg UM$ is a coproduct of L and M. Moreover $U(L \amalg M) = UL \amalg UM.\square$

Let $L \in \mathrm{ob}\,\mathbf{DL}$, and $z \in ZL_n$ (as usual the cycles of L_n). Let a be an element of degree $n+1$ and $\mathbb{L}(a)$ the free Lie algebra generated by a; define on the coproduct $L \amalg \mathbb{L}(a)$ a differential D such that $Da = z$. Then the morphism $L \subset L \amalg \mathbb{L}(a)$ is called an *elementary cofibration* in **DL**.

A map $i: L \longrightarrow M$ in **DL** is a *cofibration* if and only if there is a free subspace V of M and an isomorphism of algebras $M \cong i(L) \oplus \mathbb{L}(V)$.

Let $i: L \longrightarrow (L \amalg \mathbb{L}(V), d)$ be a cofibration and $f: L \longrightarrow K$ any map in **DL**. Then we define a cocartesian diagram

$$
\begin{array}{ccc}
L & \xrightarrow{\ f\ } & K \\
{\scriptstyle i}\downarrow & & \downarrow{\scriptstyle \bar{i}} \\
L \amalg (L(V), d) = M & \xrightarrow[\ \bar{f}\ ]{} & M \cup_L K = (K \amalg \mathbb{L}(V), d)
\end{array}
$$

by $\bar{f} = f \amalg T(V)$. The differential on $K \amalg \mathbb{L}(V)$ is defined by requiring $\bar{f}$ and $\bar{i}$ to be differential morphisms.

4.3 Acyclicity.

4.3.1 Proposition. The differential Lie algebra $(\mathbb{L}(V \oplus dV), d)$, where dV is defined by $(dV)_n = V_{n+1}$ and the differential $d \colon V \longrightarrow dV$ is defined by the shift of degree -1 on V and $d(dV) = 0$, is acyclic.

Proof. The result is an immediate consequence of the analogue one for **DA** (§3.4): just apply the following theorem of Quillen [44] and the remark that $U\mathbb{L}(V) \cong T(V)$. □

4.3.2 Theorem. The homology functor commutes with the universal enveloping algebra functor, i.e. $HUL = UHL$ for any differential Lie algebra L. □

4.4 Cofibration category.

4.4.1 Theorem. The category **DA** is a cofibration category in which all objects are fibrant.

Proof Cf. [8], chapter 1. □

4.4.2 Cofibrant models. Once again, we point out that $*$-cofibrant objects in **DL** are the free objects. ($* = 0$ is an obvious initial object in **DL**)

4.5 A cylinder in the category DL.

4.5.1 Construction of $I_B A$. Let $B \rightarrowtail A = B \amalg \mathbb{L}(V)$ be a cofibration in **DL**. Then we define $I_B A = A \amalg (\mathbb{L}(sV, dsV), d)$ where sV is a vector space isomorphic to V with degrees shifted by $+1$ and $(\mathbb{L}(sV, dsV), d)$ is the corresponding canonical acyclic Lie algebra (cf. Proposition 4.3.1).

To define the factorization

$$A \cup_B A = B \amalg \mathbb{L}(V' \oplus V'') \overset{(i_0, i_1)}{\rightarrowtail} A \amalg (\mathbb{L}(sV, dsV), d) \overset{p}{\longrightarrow} A.$$

we choose $p = (A, 0)$, and let i_0 be the canonical inclusion

$$A \subset A \amalg (\mathbb{L}(sV, dsV), d).$$

For the definition of i_1 we need a few preparations. Let S be the derivation of degree $+1$ defined by

$$S(B) = S(sV) = S(dsV) = 0$$
$$S(v) = sv.$$

Then $\theta = dS + Sd$ is a derivation of degree 0 on $I_B A$. One can verify that for each $x \in I_B A$, $\theta^N(x) = 0$ for some N. Then $e^\theta = \sum_{N=0}^{\infty} \frac{\theta^N}{N!}$ is well-defined, it is an isomorphism and commutes with d.

We define $i_1 = e^\theta i_0$. The verification that (i_0, i_1) is a cofibration is left to the reader (cf. [8]).

4.5.2 Application: Homotopy groups in DL. We now consider the analogue in the category **DL** of the sphere in **Top** (we shall say in chapter 5 that $L(\mathbb{Q}, n)$ is a model for the rational $n + 1$-sphere).

Let V be a $\mathbb{Q}$-vector space, and denote by $\underline{V}$ the graded vector space concentrated in degree 0 with $\underline{V}_0 = V$. Let $L(\underline{V}, n) = \mathbb{L}(s^n \underline{V}, d)$ be the free graded Lie algebra generated by V concentrated in degree n, together with the zero differential.

4.5.3 Proposition. For any $\Lambda \in \mathbf{DL}$, there is an isomorphism

$$[L(\mathbb{Q}, n), \Lambda] = H_n \Lambda$$

and this isomorphism is natural in Λ.

Proof. Let us consider the differential Lie algebra

$$IL(\mathbb{Q}, n) = (\mathbb{L}(\mathbb{Q} x' \oplus \mathbb{Q} x'' \oplus \mathbb{Q} sx), d)$$

with $|x| = |x'| = n$, $|sx| = n + 1$

$$dx' = dx'' = 0, \; dsx = x'' - x'.$$

One easily checks that $IL(\mathbb{Q}, n)$ is a cylinder object for $L(\mathbb{Q}, n)$.
Looking at the image of $1 \in \mathbb{Q}$ proves the natural isomorphism

$$\mathbf{DL}(L(\mathbb{Q}, n), \Lambda) \cong Z_n \Lambda$$

and moreover two morphisms $f, g: L(\mathbb{Q}, n) \longrightarrow \Lambda$ define a map

$$f \vee g: \mathbb{L}(\mathbb{Q} x' \oplus \mathbb{Q} x'') \longrightarrow \Lambda$$

which extends to $IL(\mathbb{Q}, n)$ if and only if the corresponding cycles in $Z_n \Lambda$ are homologous. □

Remark. As the reader already noticed, we did not use the cylinder we introduced in 4.5.1 because in the proof above this definition was less easy to handle. Of course corollary 3.6.2, chapter 3, justifies the use of this alternative cylinder.

Similarly, in **CDA** we obtain

4.5.4 Proposition. For any $A \in \mathbf{CDA}$ there is an isomorphism

$$[\Lambda(\mathbb{Q}, n), A] = H_n A$$

and this isomorphism is natural in A.   □

Chapter 5
The Rational Homotopy Category
of Simply Connected Spaces

Rational homotopy theory is homotopy theory up to rational homotopy equivalence. In this chapter we show that rational homotopy of 1-connected spaces fits into the general framework introduced in the previous chapter. We show how the categories $\mathbf{CDA}_*$ and $\mathbf{DL}$ introduced in chapter 4 model the category of 1-connected rational CW-spaces; more precisely, the homotopy categories of $\mathbf{CDA}_*$ and $\mathbf{DL}$ respectively, are equivalent to the homotopy category of 1-connected rational CW-spaces.

§1 The category of rational spaces

The following result is the starting point of rational homotopy theory.

1.1 Proposition and definition (Serre). A map $f : X \longrightarrow Y$ between 1-connected spaces is a rational equivalence if one of the following equivalent conditions holds
i) $f_* : H_*(X;\mathbb{Q}) \longrightarrow H_*(Y;\mathbb{Q})$ is an isomorphism of graded $\mathbb{Q}$-vector spaces;
ii) $\pi_*(f) \otimes \mathbb{Q} : \pi_*(X) \otimes \mathbb{Q} \longrightarrow \pi_*(Y) \otimes \mathbb{Q}$ is an isomorphism of graded $\mathbb{Q}$-vector spaces.

Proof. The equivalence of i) and ii) is Serre's Whitehead's theorem modulo $\mathbf{C}$, where $\mathbf{C}$ is the class of torsion abelian groups ([46]).   □

Rational equivalences are the equivalences of a suitable cofibration category.

1.2 Proposition. Let $\mathbf{CW}_\mathbb{Q}^*(\pi_1 = 0)$ be the full subcategory of $\mathbf{Top}_*$ whose objects are 1-connected CW-spaces with base point. Let us define weak equivalences to be rational equivalences and cofibrations to be ordinary cofibrations in $\mathbf{Top}$ namely maps which satisfy the homotopy extension property. Then $\mathbf{CW}_\mathbb{Q}^*(\pi_1 = 0)$ together with these weak equivalences and cofibrations is a cofibration category.

Proof. Axiom (C_1) obviously holds.
We shall prove axiom (C_2). So suppose given a cofibration $B \rightarrowtail A$ and a map $B \longrightarrow Y$. We know that $\mathbf{Top}_{\mathbf{CW}}\cdot$ is a cofibration category (cf. §3, Table 1, p. 36). Let $A \cup_B Y$ be the push-out in $\mathbf{Top}_{\mathbf{CW}}\cdot$ (then $Y \longrightarrow A \cup_B Y$ is a cofibration in $\mathbf{Top}_{\mathbf{CW}}\cdot$). This is our candidate for the push-out in $\mathbf{CW}_\mathbb{Q}^*(\pi_1 = 0)$.
By the van Kampen theorem $A \cup_B Y$ is 1-connected. Note that by excision $H_*(A, B) \cong H_*(A \cup_B Y, Y)$. Now if $B \longrightarrow A$ is a weak equivalence, then $H_*(A, B) \otimes \mathbb{Q} = 0$ and the same holds for $H_*(A \cup_B Y, Y) \otimes \mathbb{Q}$, proving that $Y \longrightarrow A \cup_B Y$ is a weak equivalence.
Mutatis mutandis, we prove that if $B \longrightarrow Y$ is a weak equivalence, then so is $A \longrightarrow A \cup_B Y$. This concludes the proof of (C_2).
The mapping cylinder construction shows that (C_3) holds in $\mathbf{CW}_\mathbb{Q}^*(\pi_1 = 0)$.
For the proof of (C_4) we use the following two lemmas.

1.2.1 Lemma. Let X be a 1-connected CW-space. Then, the reduced homology $\tilde{H}_*(X;\mathbb{Z})$ is a rational vector space if and only if $\pi_*(X)$ is rational. In this case we call X a *rational space*.

Sketch of proof. First observe that the lemma is true for $K(\mathbb{Q},n)$: in fact

$$\pi_1(K(\mathbb{Q},n)) = \begin{cases} \mathbb{Q} & \text{if } i = n, \\ 0 & \text{if } i \neq n, \end{cases} \quad \text{while}$$

$$\tilde{H}_i(K(\mathbb{Q},2m);\mathbb{Z}) = \begin{cases} \mathbb{Q} & \text{if } i = 2km, \ k > 0 \\ 0 & \text{otherwise} \end{cases} \quad \text{and}$$

$$\tilde{H}_i(K(\mathbb{Q},2m-1);\mathbb{Z}) = \begin{cases} \mathbb{Q} & \text{if } i = 2m-1, \\ 0 & \text{otherwise.} \end{cases}$$

Actually, $H_*(K(\mathbb{Q},n);\mathbb{Z}) = \Lambda_{\mathbb{Q}}(x_n)$ is the free (graded) commutative $\mathbb{Q}$-algebra on one generator of degree n: this follows by induction on n from the multiplicative property of the Serre spectral sequence applied to the fibration sequence of H-spaces $K(\mathbb{Q},n) \longrightarrow * \longrightarrow k(\mathbb{Q},n+1)$, starting with $H_*(K(\mathbb{Q},1);\mathbb{Z}) = H_*(S_{\mathbb{Q}};\mathbb{Z}) = \Lambda_{\mathbb{Q}}(x_1)$.

Now for an arbitrary 1-connected space, the lemma easily follows by induction on the Postnikov decomposition of X for the "only if" statement and the homology decomposition for the "if" statement. $\qquad\qquad \square$

1.2.2 Lemma. For each 1-connected CW-space X there exists a weak equivalence

$$q : X \xrightarrow{\sim} X_{\mathbb{Q}}$$

where $X_{\mathbb{Q}}$ is a rational space.

Proof. We do this inductively via the homology decomposition of X. For a Moore space $M(A,n)$, we set $M(A,n)_{\mathbb{Q}} = M(A \otimes \mathbb{Q}, n)$ and

$$q : M(A,n) \xrightarrow{\sim} M(A \otimes \mathbb{Q}, n) = M(A,n)_{\mathbb{Q}}$$

is induced by the natural map $A \longrightarrow A \otimes \mathbb{Q}$.

Now assume $n \geq 3$ and consider the diagram

$$
\begin{array}{ccc}
M(H_nX, n-1) & \xrightarrow{\ q\ } & M(H_nX \otimes \mathbb{Q}, n-1) \\
{\scriptstyle k_n'}\big\downarrow & & \big\downarrow{\scriptstyle (k_n')_{\mathbb{Q}}} \\
X_{(n-1)} & \xrightarrow[q_{(n-1)}]{\ \sim\ } & X_{(n-1)\mathbb{Q}} \\
\big\downarrow & & \big\downarrow \\
X_{(n)} & \xrightarrow[q_{(n-1)}]{} & X_{(n)\mathbb{Q}}
\end{array}
$$

where we assume that $X_{(n-1)_Q}$ has already been constructed.

Recall that $X_{(n)}$ and $X_{(n)_Q}$ are push-outs and let us rewrite the preceeding diagrams

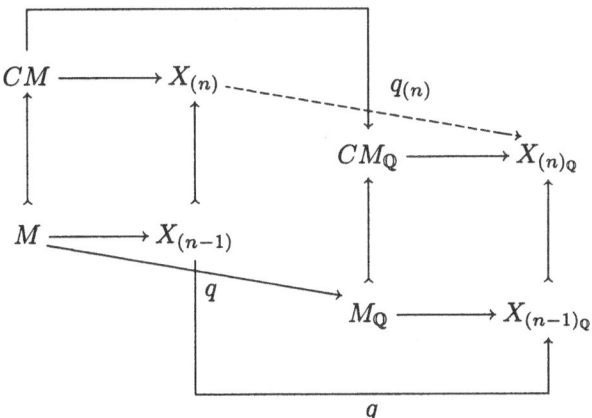

The composite $M \longrightarrow CM \longrightarrow CM_Q$ (resp. $M \longrightarrow X_{(n-1)} \longrightarrow X_{(n-1)_Q}$) is homotopic to the composite $M \longrightarrow M_Q \longrightarrow CM_Q$ (resp. $M \longrightarrow M_Q \longrightarrow X_{(n-1)_Q}$). Then there exists $q_{(n)} : X_{(n)} \longrightarrow X_{(n)_Q}$ such that the whole diagram is homotopy commutative (use the h. e. p. of Chapter 1, §2.1, for $M \rightarrowtail CM$ and an induction for $q_{(n-1)}$). The five lemma shows that $q_{(n)}$ induces an isomorphism in mod $\mathbb{Q}$-homology. Now $q : X \longrightarrow X_Q$ is the limit of the $q_{(n)}$'s.

Notice that by replacing q by its mapping cylinder we may assume that q is a cofibrant object in $\mathbf{CW}_{\mathbb{Q}}^*(\pi_1 = 0)$. $\qquad\qquad\square$

1.2.3 Lemma. A simply connected rational space is fibrant.

Proof. Let X be rational and let $f : X \xrightarrow{\sim} Y$ be a trivial cofibration. Let $q : Y \xrightarrow{\sim} Y_Q$ a trivial cofibration as above. By Theorem 2.4, Chapter 2, $q \circ f : X \rightarrowtail Y_Q$ is both a homotopy equivalence and a cofibration in **Top** which therefore admits a retraction g. Then $g \circ q$ is a retraction of f.

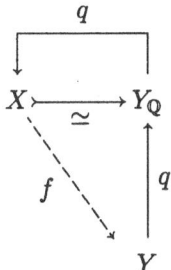

$\qquad\qquad\square$

Axiom (C_4) clearly follows from lemmas 1.2.1 and 1.2.2. This completes the proof of proposition 1.2.   □

Remark. Again by Theorem 2.4, Chapter 2, $f : X \longrightarrow Y$ is a rational equivalence if and only if $f_\mathbb{Q} : X_\mathbb{Q} \longrightarrow Y_\mathbb{Q}$ is a homotopy equivalence. Notice that the map $q : X \xrightarrow{\sim} X_\mathbb{Q}$ induces an isomorphism

$$\pi_*(X) \otimes_\mathbb{Z} \mathbb{Q} \xrightarrow{\cong} \pi_*(X_\mathbb{Q}) \otimes_\mathbb{Z} \mathbb{Q} = \pi_*(X_\mathbb{Q}).$$

We have just proved that rational spaces are exactly the fibrant and $*$-cofibrant objects in $\mathbf{CW}^*_\mathbb{Q}(\pi_1 = 0)$. Therefore, by Theorem 5.2, Chap. 3, the following theorem holds:

1.3 Theorem. The homotopy category of rational 1-connected CW-spaces is equivalent to the category $\mathrm{Ho}(\mathbf{CW}^*_\mathbb{Q}(\pi_1 = 0))$.   □

We now outline constructions for the Quillen and Sullivan minimal models, which respectively reflect the homology and homotopy decompositions.

§2   Quillen's model category

We first fix some notations.

$\mathbf{DL}(H_0 = 0)$ is the subcategory of $\mathbf{DL}$ (cf. Chapter 4, Definition 4.1) whose objects have trivial 0-homology.

$\mathbb{Q}$-**spaces**$_1$ is the category of 1-connected well-pointed rational CW-spaces. It is not difficult to see that it is a cofibration subcategory of $\mathbf{Top_{CW^*}}$.

2.1 Theorem of Quillen. There is a functor

$$\lambda : \mathbf{CW}^*_\mathbb{Q}(\pi_1 = 0) \longrightarrow \mathbf{DL}(H_0 = 0) \quad \text{(coefficients } \mathbb{R} = \mathbb{Q})$$

which carries weak equivalences to weak equivalences and homotopy push-outs to homotopy push-outs. Moreover, λ induces equivalences of categories

$$
\begin{array}{ccc}
\mathrm{Ho}\,\mathbf{CW}^*_\mathbb{Q}(\pi_1 = 0) & \xdashrightarrow[\sim]{\mathrm{Ho}\,\lambda} & \mathrm{Ho}\,\mathbf{DL}(H_* = 0) \\[2mm]
\Big\uparrow{\scriptstyle\sim} & & \Big\uparrow{\scriptstyle\sim} \\[2mm]
\mathbb{Q}-\mathbf{spaces}_1 / \simeq & \xrightarrow[\;L\;]{\sim} & \mathbf{DL}_c(H_* = 0)/\simeq
\end{array}
$$

Proof. The construction of the functor λ is quite involved. We refer to [44].   □

2.2 Definition For each 1-connected space X, $L(X)$ is called the *Quillen model* of X. Moreover, there are weak equivalences in **DL**

$$L(V, n-1) \xrightarrow{\sim} \lambda M(V, n).$$

Here $L(V, n-1) = (\mathbb{L}(V_{n-1}), d = 0)$ is the free chain Lie algebra generated by the rational vector space V in degree $n-1$, together with the trivial differential. In particular,

$$L(\mathbb{Q}, n-1) \xrightarrow{\sim} \lambda(S_{\mathbb{Q}}^n) \xleftarrow{\sim} \lambda(S^n).$$

Recalling Proposition 4.5.3, Chapter 4, we see that λ induces the isomorphism

$$\pi_n(X_{\mathbb{Q}}) = [S^n, X_{\mathbb{Q}}] \underset{\mathrm{Ho}(\lambda)}{\cong} [\lambda S^n, \lambda X_{\mathbb{Q}}] \cong H_{n-1}(\lambda X_{\mathbb{Q}})^{(3)}$$

2.3 Theorem. For each $X \in \mathbf{CW}_{\mathbb{Q}}^*(\pi_1 = 0)$, its Quillen model $L(X) \in \mathbf{DL}$ satisfies

0) $L(X)$ is free.
i) $L(X) \xrightarrow{\sim} \lambda(X)$.
ii) $H_*L(X) \cong \pi_*(\Omega X) \otimes \mathbb{Q} \cong s^{-1}\pi_*(X) \otimes \mathbb{Q}$.
iii) $L(X)/[L(X), L(X)] \cong s^{-1}H_*(X; \mathbb{Q})$.   □

Construction of $L(X)$. Replacing X by $X_{\mathbb{Q}}$, we may assume without loss of generality, that X is rational. We write for short $\tilde{H}_i = \tilde{H}_i(X)$.

We set $L_2(X) = L(H_2, 1) \xrightarrow{\sim} \lambda X_{(2)}$. Assume $L_n(X) \xrightarrow{\sim} \lambda X_{(n)}$ has been constructed. Then we define a **DL** map $d_{n+1} : L(H_{n+1}, n-1) \longrightarrow L_n(X)$ as follows: Quillen's functor λ takes the push-out diagram

$$
\begin{array}{ccc}
* & \longrightarrow & X_{(n+1)} \\
\uparrow & & \uparrow \\
& \text{p.o.} & \\
\uparrow & & \uparrow \\
M(H_{n+1}, n) & \underset{k'_{n+1}}{\longrightarrow} & X_{(n)}
\end{array}
$$

into the push-out diagram

$$
\begin{array}{ccc}
\lambda* \sim * & \longrightarrow & \lambda X_{(n+1)} \\
\uparrow & & \uparrow \\
& \text{p.o.} & \\
\uparrow & & \uparrow \\
\lambda M(H_{n+1}, n) & \underset{k'_{n+1}}{\longrightarrow} & \lambda X_{(n)}
\end{array}
$$

---

(3) This is the proof of assertion ii) of the following theorem!

Now $L(H_{n+1}, n-1) \xrightarrow{\sim} \lambda M(H_{n+1}, n)$ and by induction hypothesis $L_n(X) \xrightarrow{\sim} \lambda X_{(n)}$.

Since $L(H_{n+1}, n-1)$ is fibrant, the lifting lemma (§3.3, Chapter 3) shows the existence of $d_{n+1} : L(H_{n+1}, n-1) \longrightarrow L_n(X)$ filling in the homotopy commutative square

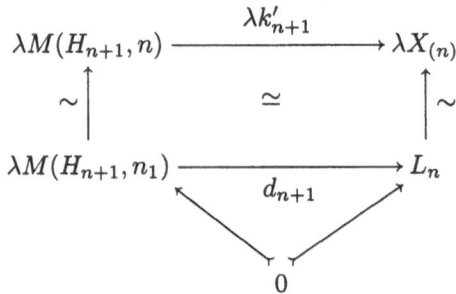

We define $L_{n+1}(X)$ by the push-out diagram

$$
\begin{array}{ccc}
CL(H_{n+1}, n) & \longrightarrow & L_{n+1} \\
\uparrow & \text{p.o.} & \uparrow \\
L(H_{n+1}, n-1) & \xrightarrow{\;\;d_{n+1}\;\;} & L_n
\end{array}
$$

where $CL(V, n) = (\mathbb{L}(s^n V, s^{n+1} V), dsv = v)$ is a contractible Lie algebra by Proposition 4.3.1, Chapter 4.

Finally the glueing lemma ([8] Chapter II, Lemma 1.2) in **DL** applied to the diagram

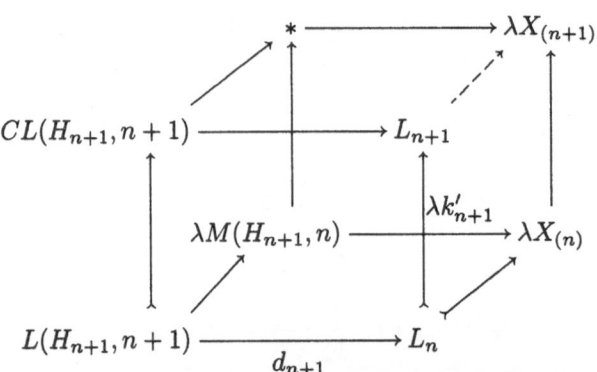

in which the front and back faces are push-outs, shows that the induced map between the push-outs is also a weak equivalence $L_{n+1}(X) \xrightarrow{\sim} \lambda X_{(n+1)}$. Finally, we set $L(X) = \varinjlim L_n(X) = (\mathbb{L}(s^{-1} H_*(X)), d)$. $\square$

Dually, one gets the minimal model of Sullivan from the Postnikov decomposition of X.

§3 Sullivan's model theory

3.1 Differential forms. The bridge between topological spaces and Sullivan models is built in two steps:

$$\mathbf{Top} \longrightarrow \mathbf{CDA}/\mathbb{Q} \longrightarrow \text{model.}$$

The second part has been described in Chapter 4, §1.6.

As to the first part, there are three constructions due to Quillen [44], Sullivan [48], and Anick [3] respectively (they all end up with the same model!). Sullivan's construction is the easiest to describe (but Anick's is probably the most suggestive) and we will just mention the ingredients:

i) For each n there is an acyclic CDA $A(n)$ of $\mathbb{Q}$-differential forms on Δ^n.
ii) For each topological space X, there is a CDA $(A_{PL}(X), d)$: an element $\Phi \in A_{PL}^p$ associates to each singular q-simplex σ (all q) an element $\Phi_\sigma \in A^p(q)$, compatible with face and degeneracy maps.
iii) Multiplication, addition and differential are then given by

$$(\Phi + \Psi)_\sigma = \Phi_\sigma + \Psi_\sigma, \quad (\Phi \wedge \Psi)_\sigma = \Phi_\sigma \wedge \Psi_\sigma, \quad (d(\Phi))_\sigma = d(\Phi_\sigma).$$

3.1.1 Definition. $(A_{PL}(X), d)$ is the CDA of piecewise linear $\mathbb{Q}$-differential forms. Its minimal model is the *Sullivan minimal model* of X, denoted by $\Lambda(X)$, and defined if $H^0(A_{PL}) = \mathbb{Q}$.

Because $(A_{PL}(X), d)$ is so closely related to the singular simplices of X it is not surprising that we have the following:

3.2 Theorem of Sullivan. There is a functor

$$A : (\mathbf{Top}_*^0)^{\mathrm{op}} \longrightarrow \mathbf{CDA}_*^0$$

where $(\mathbf{Top}_*^0)^{\mathrm{op}}$ is the fibration category of connected topological spaces with base point which carries weak equivalences to weak equivalences and certain homotopy pull-backs to certain homotopy push-outs. Moreover, A induces equivalences of categories

$$
\begin{array}{ccc}
\mathrm{Ho}\,\mathbf{CW}_{\mathbb{Q}}^*(\pi_1 = 0) & \xdashrightarrow[\sim]{\mathrm{Ho}\,A} & \mathrm{Ho}\,\mathbf{CDA}_*^0 \\
\uparrow{\scriptstyle\sim} & & \uparrow{\scriptstyle\sim} \\
\mathbb{Q}\text{-spaces of finite type}_1/\simeq & \xrightarrow[\Lambda]{\sim} & (\mathbf{CDA}_*^0 \text{ of finite type})_c(H_1 = 0)/\simeq
\end{array}
$$

Moreover, there are weak equivalences

$$\Lambda(V^{\#}, n) \xrightarrow{\sim} A(K(V, n)).$$

Here $\Lambda(W, n) = (\Lambda(W^n), d = 0)$ is the free CDA with $d = 0$ generated by W in degree n, and $V^{\#} = \text{Hom}(V, \mathbb{Q})$.

Therefore A yields isomorphisms

$$H_n(X, \mathbb{Q}) = [X, K(\mathbb{Q}, n)] \xrightarrow[\cong]{A} [AK(\mathbb{Q}, n), AX] = [\Lambda(\mathbb{Q}, n), AX] = H_n AX$$

where the last isomorphism comes from Proposition 4.5.4, Chapter 4.

The cofibrant objects of $\mathbf{CDA}_*^0(H^1 = 0)$ are the minimal models of $\mathbf{CDA}_*^0(H^1 = 0)$. In particular, the theorem above says

3.2.1 Theorem. Let $(\Lambda V, d)$ be a minimal Sullivan model on $\mathbb{Q}$ such that $V^1 = 0$ and V has finite type. Then $(\Lambda V, d)$ is the minimal model of a simply connected CW-complex of finite type.

We summarize these results in the following statement.

3.2.2 Theorem. The homotopy types of rational spaces are in one-to-one correspondance with the isomorphism classes of models.   □

We now specify some properties of the Sullivan model; on the way we outline a construction for the Sullivan minimal model which reflects the homotopy decomposition.

3.2.3 Theorem. For each $X \in \mathbf{CW}_{\mathbb{Q}}^*(\pi_1 = 0)$, its Sullivan model $\Lambda(X) \in \mathbf{CDA}$ satisfies
i) $\Lambda(X) \xrightarrow{\sim} A(X)$.
ii) $H_* \Lambda(X) \cong H_*(X, \mathbb{Q})$.
iii) $\Lambda^+(X)/\Lambda^+(X).\Lambda^+(X) = \pi_*(X; \mathbb{Q})^{\#}$.

Proof. i) merely is the definition of a Sullivan model (cf. §1.2.1, Chapter 4) and then ii) follows from the remark above $H_n(X, \mathbb{Q}) = H_n AX$.   □

Construction of $\Lambda(X)$. Replacing X by $X_{\mathbb{Q}}$, we may assume without loss of generality, that X is rational. We write for short $\pi_i = \pi_i(X)$. We set $\Lambda_2(X) = \Lambda(\pi_2^{\#}, 2) \xrightarrow{\sim} A(X_2)$.

Assume $\Lambda_n(X) \xrightarrow{\sim} AX_n$ has been constructed. Then we define the CDA map: $d_{n+1} : \Lambda(\pi_{n+1}^{\#}, n) \longrightarrow \Lambda_n(X)$ as follows: the Sullivan's functor A takes the pullback diagram

$$
\begin{array}{ccc}
X_{n+1} & \longrightarrow & * \\
\downarrow & \simeq \text{p.b.} & \downarrow \\
X_n & \xrightarrow[k_{n+1}]{} & K(\pi_{n+1}, n)
\end{array}
$$

into the push-out diagram

$$
\begin{array}{ccc}
AX_{n+1}: \longleftarrow & & * \\
\big\uparrow & \simeq \text{ p.o.} & \big\uparrow \\
AX_n \xleftarrow{\;Ak_{n+1}\;} & & AK(\pi_{n+1}, n)
\end{array}
$$

Now $\Lambda(\pi_{n+1}^{\#}, n)$ is a model of $AK(\pi_{n+1}^{\#}, n)$ and in particular $\Lambda(\pi_{n+1}^{\#}, n) \xrightarrow{\sim} AK(\pi_{n+1}, n)$ is a quasi-isomorphism; by induction hypothesis so is $\Lambda_n(X) \xrightarrow{\sim} AX_n$.

Since $\Lambda(\pi_{n+1}^{\#}, n)$ is fibrant, the lifting lemma (3.2, Chapter 3) shows that there exists

$$
d_{n+1} : \Lambda(\pi_{n+1}, n) \longrightarrow \Lambda_n(X)
$$

making the square

$$
\begin{array}{ccc}
AX_n \xleftarrow{\;\lambda Ak_{n+1}\;} & & AK(\pi_{n+1}, n) \\
\sim \big\uparrow & \simeq & \big\uparrow \sim \\
\Lambda_n \xleftarrow{\qquad\qquad} & & \Lambda(\pi_{n+1}^{\#}, n) \\
& \searrow{\scriptstyle d_{n+1}} \quad \swarrow & \\
& k &
\end{array}
$$

homotopy commutative. We define $L_{n+1}(X)$ by the push-out diagram

$$
\begin{array}{ccc}
C\Lambda(\pi_{n+1}^{\#}, n) & \longrightarrow & \Lambda_{n+1} \\
\sim \big\uparrow & \text{p.o.} & \big\uparrow \sim \\
\Lambda(\pi_{n+1}^{\#}, n) & \xrightarrow{\;d_{n+1}\;} & \Lambda_n
\end{array}
$$

where $C\Lambda(V, n) = (\Lambda(s^n V, s^{n+1} V), dv = sv)$ is a contractible commutative algebra by Proposition 4.3.1, Chapter 4.

The glueing lemma ([8] Chapter II, Lemma 1.2) in CDA applied to the diagram

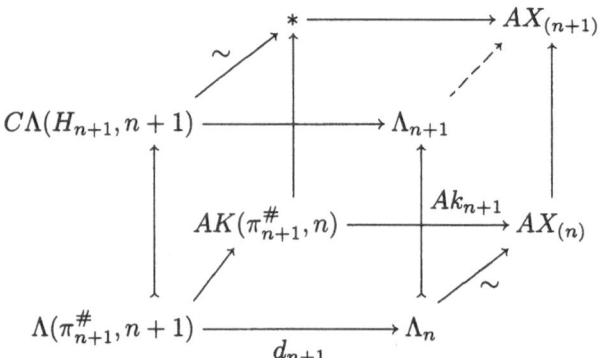

in which the front and back faces are push-outs shows that the induced map between the coproducts $\Lambda_{n+1}(X) \xrightarrow{\sim} AX_{n+1}$ is a quasi-isomorphism. Finally, we set

$$\Lambda(X) = \lim_{\longrightarrow} \Lambda_n(X) = (\Lambda(\pi_*^\#(X)), d).$$

The verification of assertion iii) is now immediate. $\square$

3.3 Examples.

Example 1. S^n.

One has

$$H^*(A_{PL}(S^n), d) = \begin{cases} H^*(S^n) \\ \mathbb{Q} & \text{if } * = 0 \text{ or } n \ . \\ 0 & \text{otherwise} \end{cases}$$

Choose an element $a \in A_{PL}(S^n)$ representing the fundamental class in $H^*(S^n)$. Then there is a map

$$\varphi : (\Lambda u, 0) \longrightarrow (A_{PL}(S^n), d)$$

where u has degree n and $\varphi(u) = a$.

a) If n is odd, then $\Lambda u = \mathbb{Q} \oplus \mathbb{Q} u$ and φ is a quasi-isomorphism such that $\varphi(u) = a$. Therefore $(\Lambda u, 0)$ is the model of S^n.

b) If n is even, then $\varphi(u^2)$ is a boundary $d\Phi$ in $A_{PL}(S^n)$. Extend φ to

$$(\Lambda(u,v), dv = u^2) \longrightarrow A_{PL}(S^n), \quad |v| = 2n - 1$$

$$v \longmapsto \Phi.$$

A $\mathbb{Q}$-linear basis of $\Lambda(u,v)$ is $\{u^r, r \geq 0; u^r v, r \geq 0\}$. Since $du^r v = u^{r+2}$, one checks that $H(\Lambda(u,v), d) = \mathbb{Q} \oplus \mathbb{Q} \bar{u}$. Therefore $\varphi : (\Lambda(u,v), d) \longrightarrow A_{PL}(S^n)$ is a quasi-isomorphism and the minimal model of S^n is $(\Lambda(u,v), d)$.
Clearly, if $n = 2k$, then $\pi_*(\Lambda u, d) = \mathbb{Q} u^*$, $|u^*| = 2k + 1$ and
if $n = 2k + 1$, then $\pi_*(\Lambda(u,v), d) = \mathbb{Q} u^* \oplus \mathbb{Q} v^*$, $|u^*| = 2k$, $|v^*| = 4k - 1$.

Theorem 3.2.3 now reads

$$\pi_*(S^{2k}) \otimes \mathbb{Q} = \begin{cases} \mathbb{Q}, & \text{if } * = 0, 2k, 4k - 1 \\ 0, & \text{otherwise;} \end{cases}$$

$$\pi_*(S^{2k+1}) \otimes \mathbb{Q} = \begin{cases} \mathbb{Q}, & \text{if } * = 0, 2k + 1 \\ 0, & \text{otherwise.} \end{cases} \quad .$$

Moreover, an easy application of Definition 1.8.3 and Remark 1.8.4 of Chapter 4 shows that the Whitehead product structure in $\pi_*(S^{2k+1}) \otimes \mathbb{Q}$ is given by $[u^*, u^*] = 2v^*$.

Example 2. $\mathbb{C}P^n$.

We know that $H^*(\mathbb{C}P^n) = \Lambda a/(a^{n+1})$, $\deg a = 2$.

We choose $z \in A^2_{PL}(\mathbb{C}P^n)$ representing the generator of the cohomology algebra. As above, we choose $\Omega \in A^{2n+1}_{PL}(\mathbb{C}P^n)$ such that $d\Omega = z^{n+1}$. We define

$$\varphi : (\Lambda(a, u), da = 0, du = a^{n+1}) \longrightarrow A_{PL}(\mathbb{C}P^n)$$

$$a \longmapsto z$$

$$u \longmapsto \Omega.$$

Again one easily checks that the map

$$\bar{\varphi} : \Lambda(a, u) \longrightarrow \Lambda a/(a^{n+1})$$

$$a \longmapsto a$$

$$u \longmapsto 0$$

is a quasi-isomorphism. Therefore $\pi^*(\mathbb{C}P^n) \otimes \mathbb{Q} = \mathbb{Q}a^* \oplus \mathbb{Q}u^*$.

Example 3. Let f, g be the two inclusions $S^2 \vee S^2$. They form the canonical basis of $\pi_2(S^2 \vee S^2)$. Let us consider $X = S^2 \vee S^2 \cup_{[f,[f,g]]} e^5$.

Clearly:

1°) $H^*(X; \mathbb{Z})$ is the free abelian group on 1, α, β (both of degree 2), γ (of degree 5) and all products are zero.

2°) Let z, z' be cycles in $A^2_{PL}(X)$ representing α, β. We start off our model with

$$\Lambda(a, b) \longmapsto A^2_{PL}(X)$$

$$a \longmapsto z$$

$$b \longmapsto z'$$

and $da = db = 0$.

3°) Since there is no cohomology in degree 4, there must exist elements Φ, Ψ and Ω in $A_{PL}(X)$ such that $z^2 = d\Phi$, $zz' = d\Psi$, $z'^2 = d\Omega$.

We introduce new variables u, v, w of degree 3 with $du = a^2$, $dv = ab$, $dw = b^2$

$$u \longmapsto \Phi, \ v \longmapsto \Psi, \ w \longmapsto \Omega.$$

4°) Now we compute $H(\Lambda(a, b, u, v, w))$:

$$H^1 = 0, \quad H^2 = \text{span}([a], [b]), \quad H^3 = 0, \quad H^4 = 0,$$
$$H^5 = \text{span}([av - bu], [bv - aw]).$$

Since $\dim H^5 = 1$, some linear combination $\lambda(av - bu) + \mu(bv - aw)$ must be a boundary. Thus we need to introduce a new variable x with $\deg x = 4$ and

$$dx = \lambda(av - bu) + \mu(bv - aw).$$

On the other hand, in π_* we have

$$a^* = f, \ b^* = f, \ v^* = [a^*, b^*] = [f, g]$$
$$\langle [a^*, [a^*, b^*]], x \rangle = \langle [a^*, v^*], x \rangle = \lambda.$$

Now $[a^*, [a^*, b^*]] = [f, [f, g]]$ is zero in $\pi_*(X)$ and $\lambda = 0$. Thus

$$dx = bv - aw$$

and $av - bu$ is not a boundary. So $[av - bu]$ must correspond to γ up to a non-zero multiplicative coefficient and our model

$$\Lambda(a, b, u, v, w, x) \longrightarrow A_{PL}(X)$$

is an isomorphism in cohomology in degrees ≤ 5.

In contrast with the preceding examples, we must go on adding variables to kill H^6, H^7 and the process never terminates. Indeed one knows that $\pi_*(S^2 \vee S^2) \otimes \mathbb{Q}$ is the suspension of the graded free Lie algebra on two generators of degree 1, which is infinite dimensional. Moreover, $\lim_{n \to \infty} \dim_{\mathbb{Q}} \pi_n(S^2 \vee S^2) \otimes \mathbb{Q} = \infty$.

Example 4. $X = S^2 \vee S^2 \vee S^5$.

This space has the same cohomology ring as in example 3, but its model is

$$\Lambda(a, b, y), da = db = dy = 0, \ \deg a = \deg b = 2, \ \deg y = 5.$$

Here the cohomology class of degree 5 is a sphere and so corresponds to a generator whereas in Example 3 the class of degree 5, namely $[av - bu]$, was quadratic.

We conclude this chapter with immediate applications of the Quillen and Sullivan models.

§4 Some easy applications

4.1 Theorem. Let X be a simply connected space such that $\tilde{H}_{\text{even}}(X;\mathbb{Q}) = 0$. Then $X_{\mathbb{Q}}$ is rationally homotopy equivalent to a wedge of spheres.

Proof. $\mathbb{L}(s^{-1}\tilde{H}_*(X;\mathbb{Q}))$ has all its non-zero elements in even degrees; since the differential d is of degree -1, it is zero. Therefore

$$\pi_*(\Omega X) \otimes \mathbb{Q} = \mathbb{L}(s^{-1}\tilde{H}_*(X;\mathbb{Q}))$$

and so is a free Lie algebra. Choosing a basis for $s^{-1}\tilde{H}_*(X;\mathbb{Q})$ defines a homotopy equivalence $\bigvee S_{\mathbb{Q}}^{n_i} \longrightarrow X_{\mathbb{Q}}$. □

4.2 "Dual" theorem. Let X be a connected simply space such that $\pi_{\text{odd}}(X) \otimes \mathbb{Q} = 0$. Then $X_{\mathbb{Q}}$ is homotopy equivalent to a product of Eilenberg-Mac Lane spaces.

Here $H_*(X,\mathbb{Q}) = \Lambda(\pi_*^{\#})$ is a free commutative algebra. We leave the details to the reader. □

4.3 Models of Moore and Eilenberg spaces.

In this paragraph we outline what we already mentionned in §§2 and 3 about the minimal Quillen (resp. Sullivan) model of a Moore (resp. Eilenberg-Mac Lane) space. These are the building blocks of the Quillen (resp. Sullivan) model of any space.

4.3.1 The Quillen model of a Moore space. Let $M(V,n)$ be the Moore space whose reduced homology is equal to the $\mathbb{Q}$-vector space V in dimension n. By Theorem 2.3 its Quillen model is a free Lie algebra generated by V in dimension $n-1$; its differential is then necessarilly zero; we denoted the model by $L(V,n-1) = (\mathbb{L}(V_{n-1}), d = 0)$. We thus recover the result of Hilton-Milnor about the Homotopy Lie algebra of a Moore space [52].

4.3.2 The Sullivan model of an Eilenberg-Mac Lane space. Let $K(V,n)$ be the Eilenberg-Mac Lane space whose homotopy is concentrated in degree n and equal to the $\mathbb{Q}$-vector space V. By Theorem 3.2.3 its Sullivan model is a free commutative algebra generated by V in dimension n; its differential is then necessarily zero; we denoted the model by $\Lambda(V,n) = (\Lambda(V^n), d = 0)$. We thus recover the result of Eilenberg-Milnor about the cohomology algebra of an Eilenberg-Mac Lane space [52].

Refering to their role in the homology (resp. homotopy) decomposition, we may call Moore (resp. Eilenberg-Mac Lane) spaces homology (resp. homotopy) 1-stage spaces. More generally, a homology (resp. homotopy) n-stage space is obtained from a $(n-1)$-stage space X as the cofibre of a map $M(H,m) \longrightarrow X$ (resp. $K(\pi,m) \longrightarrow X$). We give now some precisions about 2-stage spaces.

4.4 Rational 2-stage spaces.

Recall that a homology (resp. homotopy) two-stage space is the cofibre (resp. fibre) of some map $k' : M(H_m, m-1) \longrightarrow M(H_n, n)$ (resp. $k : K(\pi_n, n) \longrightarrow K(\pi_m, m+1)$). We assume that H_m, H_n (resp. π_m, π_n are finite dimensional rational vector spaces.

Homology two-stage spaces. Let d correspond to k' through the sequence of isomorphisms:

$$k' \in \quad [M(H_m, m-1), M(H_n, n)]$$

$$\Big\downarrow \cong$$

$$[M(H_m, m-2), \Omega M(H_n, n)]$$

$$\Big\downarrow \cong \pi_{m-2} = \text{homotopy functor}$$

$$\text{Hom}(H_m, \pi_{m-2}\Omega M(H_n, n))$$

$$\Big\downarrow \cong$$

$$\text{Hom}(H_m, \mathbb{L}(H_n^{n-1})_{m-2})$$

$$\Big\downarrow \cong$$

$$d \in \quad \text{Hom}_{-1}(s^{-1}H_m, \mathbb{L}(s^{-1}H_n))$$

We assert that d is the differential of the minimal Quillen model of $X = C_{k'}$ (the diagram only says that k' is characterized by a linear map $H_m \to \pi_{m-2}(\Omega M(H_n, n))$ which in turn gives a map between Quillen models $\mathbb{L}(M(H_m, m-1) \to M(H_n, n)$. The construction of $L(X)$ (cf. §2.3) shows that this last map is the differential). Therefore (see §2 again!) the Quillen model of $C_{k'}$ is

$(\mathbb{L}(s^{-1}H_m \oplus s^{-1}H_n), d)$.   □

Homotopy two-stage spaces. Let d correspond to k through the sequence of isomorphisms:

$$k \in \quad [K(\pi_n, n), K(\pi_m, m+1)]$$

$$\Big\downarrow \cong H^{m+1} = \text{(contravariant) cohomology functor}$$

$$\text{Hom}(\pi_m^{\#}, H^{m+1}(K(\pi_n^{\#}), n))$$

$$\Big\downarrow \cong$$

$$\text{Hom}(\pi_m^{\#}, \Lambda(\pi_n^{\#})_{m+1})$$

$$\Big\downarrow \cong$$

$$d \in \quad \text{Hom}_1(\pi_m^{\#}, \Lambda(\pi_n^{\#}))$$

We assert that d is the differential of the Sullivan minimal model of $X = P_k$ (interpretation similar to the preceding case: use §3.2.3). This means that the Sullivan model of P_k is $(\Lambda(\pi_m^{\#} \oplus \pi_m^{\#}), d)$.   □

Appendix
Relations between the various models of a space

Here we consider algebras, commutative algebras and Lie algebras over a field k of arbitrary characteristic.

A.1 A functor between DL and CDA

Consider a k-Lie algebra L and do not consider the eventual differential for the moment. The Lie bracket induces the following commutative triangle:

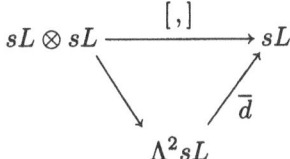

If L is of finite type this gives by dualization

$$\Lambda^2 sL \xleftarrow{d} sL^*$$

Explicitly we define the differential d by the following formulae (here $\langle\,;\rangle$ denotes the pairing of a linear form on sL):

$$\langle df; sx \wedge sy \rangle = (-1)^{|x|}, \quad \langle f; \bar{d}(sx \wedge sy) \rangle = (-1)^{|x|}, \quad \langle f; s[x \wedge y] \rangle$$

The map $L^* \longrightarrow L^*$ is of degree $+1$ and extends to a derivation on the commutative algebra (L^*, d). The property $d^2 = 0$ is just the transcription of the Jacobi identity. Now, if (L, δ) is a differential graded Lie algebra, the transposed map $\delta^* : L^* \longrightarrow L^*$ also extends to a derivation of $\Lambda L^* ((\delta^*)^2 = 0)$ which satisfies $\delta^* d + d\delta^* = 0$.

A.1.1 Definition. We define a differential commutative algebra by

$$C^*(L, \underline{d}) = (\Lambda(sL^*), d + \delta^*).$$

Remark. The dual of $C^*(L, \underline{d})$ is the *chain complex* of (L, δ) defined by

$$(C_*(L), d + \partial) = (\Lambda sL, d + \partial)$$

where

$$d(s\alpha_1 \wedge \cdots \wedge s\alpha_p) = \Sigma \pm s\alpha_1 \wedge \cdots \wedge d(s\alpha_i) \wedge \cdots \wedge s\alpha_p$$

$$\partial(s\alpha_1 \wedge \cdots \wedge s\alpha_p) = \Sigma \pm s[s\alpha_i, s\alpha_j] \wedge s\alpha_1 \wedge \cdots \wedge s\hat{\alpha}_i \wedge \cdots \wedge s\hat{\alpha}_j \wedge \cdots \wedge s\alpha_p.$$

Moreover, if L is a k-Lie algebra without differential (where k is a field), $C^*(L)$ gives rise to a canonical resolution of k as a trivial UL-module. One gets:

A.1.2 Theorem. $H_*(C^*(L)) = \text{Ext}_{UL}(k,k)$ and the word length is in bijective correspondance with the homological degree in Ext.

Proof. Cf. [20]. □

A.1.3 Proposition. If $L = \pi_*(\Lambda V, d)$ then $(C^*(L), \delta) = (\Lambda V, d_2)$.

Proof. Cf.[50]. □

Remark. L is defined from $(\Lambda V, d_1)$ (cf. Chapter 4, §1.8.2 and 1.8.3); in return L gives $(\Lambda V, d_1)$. Then the preceding proposition means that L is fully determined by $(\Lambda V, d_1)$ and vice-versa.

A.2 Models over $\mathbb{Z}/p\mathbb{Z}$

A.2.1 Adams-Hilton models. For any 1-connected CW-complex X of finite type, we can choose a CW-decomposition (there is no functorial choice). Corresponding to this decomposition, one defines an Adams-Hilton model (cf. §3.3.2, Chapter 4).

To each CW-decomposition of X we can associate a rational CW-decomposition of the rationalization X_0 (simply mimic the procedure which associates to every integral homology decomposition a rational one as in §1.2.1).

Moreover, just as we already defined the rational sphere $S_0^n = M(\mathbb{Q}, n)$, we can define the mod p–sphere $S_n^p = M(\mathbb{Z}/p, n)$ and the mod p–cell $e_p^{n+1} = CM(\mathbb{Z}/p, n)$. Then we define in the obvious way a localization of X at p, call it X_p and associate to every CW-decomposition of X a mod p–decomposition of X.

In conclusion we constructed a space X_p and a map $X \longrightarrow X_p$ such that
1) X_p is an union of p-cells, with one cell for each element of a basis of $H_*(X; \mathbb{Z}/p)$;
2) the map $X \longrightarrow X_p$ induces an isomorphism in $\mathbb{Z}/p\mathbb{Z}$-homology.

Then we can choose an Adams-Hilton model corresponding to the decomposition of X_p (resp. X_0), that is (cf. [4]):

A.2.1.1 Theorem. The *Adams-Hilton model* of X over $\mathbb{Z}/p\mathbb{Z}$ (resp. $\mathbb{Q}$) is a DA over $\mathbb{Z}/p\mathbb{Z}$ (resp. $\mathbb{Q}$) given by the following algebra $T(V) = T(v_1, v_2, \dots)$ where the v_i correspond bijectively to the cells of X_p such that

$$H_*(T(V), d) = H_*(\Omega X; \mathbb{Z}/p\mathbb{Z}) \quad (\text{resp. } H_*(\Omega X; \mathbb{Q})).$$

Moreover, there exists a weak equivalence $(T(v), d) \longrightarrow C_*(\Omega X_p; \mathbb{Z}/p\mathbb{Z})$ (resp. $C_*(\Omega X_0; \mathbb{Q})$) with $V = s^{-1}H_*(X; \mathbb{Z}/p\mathbb{Z})$ (resp. $s^{-1}H_*(X; \mathbb{Q})$) (C_* actually denotes here the cubical chain complex of ΩX, see [1] for the definition). □

Nota Bene. The Adams-Hilton model is definitely not functorial. However, if we take into account a model of the diagonal $X \longrightarrow X \times X$, there is some kind of functoriality (cf. [3]).

A.2.2 Quillen models. It is a direct consequence of an old result of Milnor and Moore [40] that if $(\mathbb{L}(V), \delta)$ is the Quillen model of a 1-connected CW-complex X, one has $HU\mathbb{L}(V) \cong H\Omega X \ (\cong UH\mathbb{L}(V))$. Anick has extended this result in the following way.

A.2.2.1. Theorem (D. Anick). If X_p is finite dimensional and $p \geq \dim X$, $(T(V), d)$, the Adams-Hilton model of X over $\mathbb{Z}/p\mathbb{Z}$, is the enveloping algebra of a certain DL $(\mathbb{L}(V), \delta)$ such that

$$(T(V), d) = U(\mathbb{L}(V), \delta) \quad \text{and}$$
$$H_*(\Omega X; \mathbb{Z}/p) = U\pi_*(\Omega X_p) = UH(\mathbb{L}(V), d).$$

Proof. Cf. [3]. □

A.3 Sullivan models

The preceding result relates Quillen models to Adams-Hilton models. We now connect Quillen and Sullivan models.

To $(\mathbb{L}(V), d)$ we apply the functor C^* (cf. definition A.1.1). We take the minimal model of the CDA:

$$\Lambda(W), d) \longrightarrow C^*(\mathbb{L}(V), d).$$

A.3.1 Theorem. The minimal model $(\Lambda(W), d)$ satisfies

$$H^*(\Lambda W, d) = H^*(X; \mathbb{Q})$$

$$(\Lambda W, d_1) = C^*(L), \text{ where } L = H_*(\mathbb{L}(V), d) = \pi_*(\Omega X) \otimes \mathbb{Q}.$$

Proof. Cf. [2]. □

Chapter 6
Attaching Cells in Topology and Algebra

A fundamental problem in algebraic topology is to understand the effect of attaching a cell to a space on the homotopy invariants of the space, such as the homotopy groups. In this chapter, we use the algebraic models of Quillen and Adams-Hilton to obtain some information on this question.

§1 Algebraic models of spaces with a cell attached

1.1 Attaching a cell to a space. Let X be a space in $\mathbf{Top}^*(\pi_1 = 0)$ and let $f : S^{n+1} \longrightarrow X$ represent a class $[f] \in \pi_{n+1}(X)$. The space $X \cup_f e^{n+2} = Y$ is defined by the push-out square

(∗)

$$
\begin{array}{ccc}
\Sigma S^n = S^{n+1} & \xrightarrow{\ f\ } & X \\
\downarrow & & \downarrow \\
e^{n+2} & \longrightarrow & X \cup_f e^{n+2} Y
\end{array}
$$

where the vertical maps are cofibrations. One says that Y is obtained from X by attaching an $(n+1)$-cell along f. The map f is called the attaching map.

We now describe the corresponding model diagrams in **DL**, **DA** and **CDA**.

1.2 Translation into Quillen models. As we saw earlier (§4.3, Chapter 5), a model of the map $f : S^{n+1} \longrightarrow X$ in the category **DL** is

$$
\varphi : (\mathbb{L}(a); 0) \longrightarrow L_X
$$

where L_X is a Quillen model for X, $\mathbb{L}(a)$ is the free Lie algebra on a single generator a of degree n and $\varphi a = z$ is a cycle in L_X whose class in HL_X corresponds to $[f]$ in the isomorphism

$$
H_n L_X = \pi_n(\Omega X) \otimes \mathbb{Q} = \pi_{n+1}(X) \otimes \mathbb{Q}.
$$

Since the Quillen functor is a model functor, a model of the diagram (∗) in **DL** is

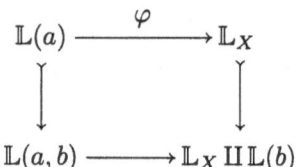

where $db = a$ in $\mathbb{L}(a, b)$ and therefore $db = \varphi a = z \in L_X \amalg \mathbb{L}(b)$.

Thus $L_X \amalg \mathbb{L}(b)$ is a model of $X \cup_f e^{n+2}$, and this model is minimal if and only if $z \in [L_X, L_X]$.

Consider now the projection

$$HL_X \longrightarrow H(L_X/[L_X, L_X]) = L_X/[L_X, L_X].$$

A more precise version of Theorem 2.3 of Chapter 5 asserts that it corresponds to the Hurewicz homomorphism $h_{\mathbb{Q}} : \pi_*(X) \otimes \mathbb{Q} \longrightarrow H_*(X; \mathbb{Q})$ under the isomorphisms (of degree -1) $\pi_*(X) \otimes \mathbb{Q} \cong HL_X$ and $\tilde{H}_*(X; \mathbb{Q}) \cong L_X/[L_X, L_X]$. Thus $z \in [L_X, L_X]$ iff $h([f] \otimes 1) = 0$ in $H_*(X; \mathbb{Q})$.

In that case, we shall freely write $L_{X \cup_f e^{n+2}} = L_X \amalg \mathbb{L}(b)$.

1.3 Translation into Adams-Hilton models. Similarly to §1.2, an Adams-Hilton model (cf. §3.3.2, Chapter 4) of $(*)$ in **DA** is

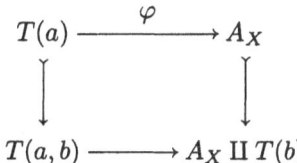

where A_X is an Adams-Hilton model of X and $\varphi(a) = z \in ZA_X$ is a cycle such that $[z]$ is the image of $[f]$ under the composition

$$\pi_{n+1}(X) = \pi_n(\Omega X) \xrightarrow{h} H_n(\Omega X) \xrightarrow{\cong} H_n(A_X),$$

h being the Hurewicz morphism.

Again $A_X \amalg T(b)$ with $db = z$ is an Adams-Hilton model of $X \cup_f e^{n+1}$, which is minimal iff z is decomposable i.e. $z \in \bar{A}_X^2$, that is iff the image $jh([f])$ is zero in $H_*(X; k)$ (here $h : \pi_*(X) \longrightarrow H_*(X; \mathbb{Z})$ is the Hurewicz homomorphism and $j : H_*(X; \mathbb{Z}) \longrightarrow H_*(X; k)$ is the canonical map).

Thus the Quillen and Adams-Hilton models of $X \cup_f e^{n+1}$ are easy to construct. We note in passing that it is much harder to give an explicit description of the Sullivan model of $X \cup_f e^{n+1}$.

1.4 Translation into Sullivan models. Indeed let $\Lambda_X = \Lambda V$ be a Sullivan model of X. Let $\alpha = [f] \otimes 1 \in \pi_{n+1}(X) \otimes \mathbb{Q}$, and assume $\alpha \neq 0$. Since $V_{\mathbb{Q}}^* = \mathrm{Hom}(\pi_*(X), \mathbb{Q})$, α defines a map $\varphi : V_{\mathbb{Q}}^* \longrightarrow \mathbb{Q}$ by $\varphi(v) = v(\alpha)$, which extends to a map in **CDA**

$$\bar{\varphi} : \Lambda V \longrightarrow (\mathbb{Q}1 \oplus \mathbb{Q}w, 0)$$

where $|w| = n + 1$, and

$$\bar{\varphi}(v) = \varphi(v).w \quad \text{if } v \in V^{n+1}$$

$$\bar{\varphi}(v) = 0 \quad \text{if } v \neq n + 1$$

$$\bar{\varphi}(1) = 1.$$

$\bar{\varphi}$ is clearly a differential map by minimality of ΛV.

Now let K be the ideal Ker $\bar{\varphi}$; then $\mathbb{Q}.1 \oplus K$ is in **CDA** and we can take its minimal model ΛW. This is a Sullivan minimal model of $X \cup_f e^{n+1}$. Notice that, if $\alpha = 0$, then $X \cup_f e^{n+1}$ is rationaly equivalent to $X \vee S^{n+1}$ and its minimal model is a minimal model of $\Lambda_X \otimes (\mathbb{Q}1 \oplus \mathbb{Q}\omega, 0)$.

In conclusion, computing $\pi_*(X \cup_f e^{n+2}) \otimes \mathbb{Q}$ amounts to compute the cohomology of the differential Lie algebra $L_X \amalg L(b)$, or equivalently the vector space W where ΛW is a minimal model of $\mathbb{Q}.1 \oplus K$.

§2 Inertia

In this section, we use the Quillen model to study the Lie algebra map

$$\pi_*(\Omega j) \otimes \mathbb{Q} : \pi_*(\Omega X) \otimes \mathbb{Q} \longrightarrow \pi_*(\Omega Y) \otimes \mathbb{Q}$$

induced by the inclusion

$$j : X \hookrightarrow X \cup_f e^{n+1} = Y$$

whose model is

$$j : L_X \hookrightarrow L_X \amalg L(b) = L_Y, \quad db = z \in L_X.$$

Recall that we have the commutative diagram of graded Lie algebras:

$$
\begin{array}{ccc}
HL_X & \xrightarrow{\;\;\;H(j)\;\;\;} & H(L_X \amalg L(b)) \\[2pt]
\cong \Big\downarrow & & \Big\downarrow \cong \\[4pt]
\pi_*(\Omega X \otimes \mathbb{Q}) & \xrightarrow{\;\;\pi_*(\Omega j \otimes \mathbb{Q})\;\;} & \pi_*(\Omega Y \otimes \mathbb{Q})
\end{array}
$$

Clearly, $H(j)$ factorizes as

(2.0) $HL_X \twoheadrightarrow HL_X/([z]) \xrightarrow{\;\gamma\;} HL_Y.$

Question: When is γ injective or surjective?

2.1 Examples. Here are three simple but standard examples:

Example 1. Let us take $X = S^2$, $f : S^3 \longrightarrow S^2$ the Hopf map; then $Y = S^2 \cup_f e^4$ is the complex projective plane $\mathbb{C}P(2)$.

The factorisation (2.0) reads:

$$
\begin{array}{ccc}
L(a) & \xrightarrow{\hspace{3cm}} & H(\mathbb{L}(a,b), db = [a,a]) = \pi(\Omega \mathbb{C}P(2)) \otimes \mathbb{Q} \\[6pt]
 & \searrow \qquad \nearrow & \\[6pt]
 & L(a)/[a,a] = \mathbb{Q}<a> &
\end{array}
$$

From the homotopy exact sequence of the Hopf fibration

$$S^1 \longrightarrow S^3 \longrightarrow \mathbb{C}P(2)$$

we get $\pi_2(\mathbb{C}P(2)) \otimes \mathbb{Q} = \mathbb{Q} = \pi_5(\mathbb{C}P(2)) \otimes \mathbb{Q}$ and $\pi_i(\mathbb{C}P(2)) \otimes \mathbb{Q} = 0$ for $i \neq 2, 5$.

Hence: $\pi_*(\Omega \mathbb{C}P(2) \otimes \mathbb{Q} = \mathbb{Q}\, a_1 \oplus \mathbb{Q} w_4$ with the trivial Lie bracket and so the map $L(a)/[a,a] \longrightarrow H(\mathbb{L}(a,b),d)$ is injective. Notice that w corresponds to the class of the cycle $[a,b] \in \mathbb{L}(a,b)$.

Example 2. $S^3 \times S^3 = S^3 \vee S^3 \cup e^6$, the attaching map is by definition the universal Whitehead product map (cf. Chapter 2, §2.6.6)

$$[\,,]: S^5 \longrightarrow S^3 \vee S^3.$$

We translate the fundamental diagram using the Quillen models.

The model of $S^3 \times S^3$ is: $(\mathbb{L}(a,b,c), dc = [a,b])$.

The model of $S^3 \vee S^3$ is: $(\mathbb{L}(a,b), d = 0)$.

Now

$$H(\mathbb{L}(a,b,c),d) = \pi_*(S^3 \times S^3) \otimes \mathbb{Q}$$

$$= (\pi(S^3) \otimes \mathbb{Q}) \times (\pi(S^3) \otimes \mathbb{Q})$$

$$= \mathbb{L}(a,b)/[a,b])).$$

In that case γ is an isomorphism.

More generally one may attach several cells at once, namely consider an attaching map

$$f : \bigvee S^{n_i+1} \longrightarrow X$$

and its maping cone $Y = X \cup (\bigvee e^{n_i+2})$.

Then the Quillen model of Y is

$$L_X \amalg \mathbb{L}(b_i)$$

with $|b_i| = n_i + 1$ and $db_i = z_i$ is a cycle in L_X whose class represents the i-th component $f_i : S^{n_i+1} \longrightarrow X$.

Again the map $HL_X \longrightarrow H(L_X \amalg \mathbb{L}(b_i))$ induced by the inclusion factors as $\gamma \pi$ where $\pi : HL_X \longrightarrow\!\!\!\longrightarrow HL_X/(\bar{z}_i)$ is the canonical surjection.

Example 3. The 6-skeleton of $S^3 \times S^3 \times S^3$ is the subset

$$T_1(S^3, S^3, S^3) = \{(x_1, x_2, x_3); \exists i, x_i = *\}$$

(the "fat wedge"). Since it is the union of three products $S_i^3 \times S_j^3$, $1 \leq i < j \leq 3$, its Quillen model is

$$\mathbb{L}(a_1, a_2, a_3, c_1, c_2, c_3), |a_i| = 2, |c_i| = 5, dc_i = [a_j, a_k]$$

with (i,j,k) a circular permutation of (1, 2, 3).

Then $L_{S^3 \times S^3 \times S^3} = L_{T_1} \amalg \mathbb{L}(e)$ with $|e| = 8$ and $de = [a_1, c_1] + [a_2, c_2] + [a_3, c_3]$, and $\pi_*(T_1) = H(L_{T_1}) = \langle a_1, a_2, a_3 \rangle \amalg \mathbb{L}(w)$, where $w = [a_1, c_1] + [a_2, c_2] + [a_3, c_3]$.

So

$$\gamma : H(L_{T_1}/(w)) \longrightarrow H(L_{S^3 \times S^3 \times S^3}) = \pi_*(\Omega(S^3 \times S^3 \times S^3)) = \langle a_1, a_2, a_3 \rangle$$

is an isomorphism.

2.2 The case where X is a wedge of spheres. Much information can be obtained on $H(L_X \amalg L(b_i))$ when X is a wedge of spheres, i.e. when L_X is free with the trivial differential.

Let $L_X = \mathbb{L}(a_i)$, $da_i = 0$ and let $db_j = z_j \in \mathbb{L}(a_i)$. Let $A = \mathbb{L}(a_i)/z_j$ and $B = H(\mathbb{L}(a_i, b_i)) = B$. The sequence (2.0) reads:

$$H(\mathbb{L}(a_i)) = \mathbb{L}(a_i) \twoheadrightarrow A \xrightarrow{\gamma} B$$

The key observation in this case is that γ admits a retraction of graded Lie algebras; indeed the map $\beta : \mathbb{L}(a_i, b_j) \longrightarrow \mathbb{L}(a_i)/l_j = A$ defined by $\beta b_j = 0$, $\beta a_j = a_i$ is a differential map and clearly $H(\beta) \circ \gamma = A$.

Let K be the kernel of $H(\beta)$. There is a short exact sequence of Lie algebras

$$0 \longrightarrow K \longrightarrow B \xrightarrow{H(\beta)} A \longrightarrow 0$$

and since γ is a section of $H(\beta)$, B appears as a semi-direct product of K by A.

On the other hand the presentation of A by generators $\{a_i\}_i$ and relations $\{l_j\}_j$ gives rise to a partial free resolution of $\mathbb{Q}$ as a trivial UA-module:

$$UA \otimes \langle l_j \rangle_j \longrightarrow UA \otimes \langle a_i \rangle_i \longrightarrow UA \longrightarrow \mathbb{Q}.$$

Let $N_3 = \ker(UA \otimes \langle l_j \rangle_j \longrightarrow UA \otimes \langle a_i \rangle_i)$. This UA-module is called the third syzygy module associated to the partial resolution. As usual we denote by sN_3 the suspension of N_3.

2.2.1 Theorem.
1) $K = \mathbb{L}(W)$
2) W is an UA-module and $W = sN^3$. B acts naturally on QK not on W (although $QK \cong W$).

The theorem is a direct application of the following result.

2.2.2 Theorem (Baues-Félix-Thomas-Lemaire). Let R be a principal ideal domain and let V_i, $i = 0, 1$, denote a R-module, the elements of which are assumed to be of degree i. Suppose we are given a free differential algebra $(T(V_0 \oplus V_1), d)$ with $dV_1 \subset T(V_0)$. Using the following notations

$$H_* = H(T(V_0 \oplus V_1), d),$$

$$I = (dV_1) \subset T(V_0),$$

$$J = \mathrm{Ker}((T(V_0 \oplus V_1) \longrightarrow T(V_0)/I = H_0),$$

we get the following isomorphisms:
a) $H_* = T_{H_0}(H_1)$;
b) $H_1 = J/J^2$;
c) H_1 is naturally isomorphic to the third syzygy module in a resolution of H_1 as a H_1-bimodule naturally associated to the differential algebra

$$(T(V_0 \oplus V_1), d).$$

Proof. See [10] and [26]. □

Proof of Theorem 2.2.1. Endow the Lie algebra $\mathbb{L}(a_i, b_i)$ (resp. $\mathbb{L}(a_i)/z_j$) with a second abstract degree defined by $\deg a_i = 0 = \deg z_j$ and $\deg b_i = 1$.

Now apply Theorem 2.2.2 to the free associative algebra $U\mathbb{L}(a_i, b_j)$ where V_0 is generated by the a_i and V_1 by the b_j and use the notations of this theorem. In this case J is the enveloping algebra of the kernel of the obvious map $\mathbb{L}(a_i, b_j) \to A$. As a sub-algebra of a free Lie algebra, this kernel is a free Lie algebra and therefore J is a free associative algebra generated by some $\mathbb{Q}$-vector space W.

Now the differential on J^2 is zero and then $UK = J$. Because $J/J^2 = W$ Theorem 2.2.2 ends the proof. □

We conclude this paragraph by setting the preceding considerations in a somewhat more general framework: that is the notion of inert element (here X is no longer supposed to be a wedge of spheres; on the other hand we attach only one sphere at a time).

2.3 Inert elements. Consider the following attaching map

$$S^{n+1} \xrightarrow{f} X \longrightarrow X \cup_f e^{n+2} = Y$$

which gives the diagram

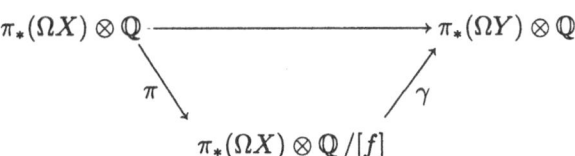

2.3.1 Theorem. The following statements are equivalent
i) γ is surjective.
ii) γ is an isomorphism.
iii) $L = \ker \pi$ is free and QL is UA-free.
iv) $U\pi$ induces an injection on $\mathrm{Tor}_2(\mathbb{Q}, \mathbb{Q})$ and an isomorphism on $\mathrm{Tor}_3(\mathbb{Q}, \mathbb{Q})$.
v)   The homotopy fibre of f is a wedge of spheres.

Proof. [32]. □

2.3.2 Definition. Such an element $[f] \in \pi_*(\Omega X) \otimes \mathbb{Q}$ is called *inert*.

Example. $Y = S^3 \times S^3 \times S^3 = X \cup_w e^9$ where X is the 6-skeleton.

$$\pi_*(\Omega X) \otimes \mathbb{Q} = \langle a, b, c \rangle \amalg \mathbb{L}(w)$$

and w is inert.

The notion af inertia leads to a beautiful application in the case of *Poincaré duality complexes.*

Let V be a Poincaré complexe on $\mathbb{Q}$ such that $\pi_1(V) = 0$. Suppose V is of dimension N and let V' be the $(N-2)$-skeleton:

$$V = V' \cup_w e^{N-2}.$$

2.3.3 Theorem. w is inert in $\pi_*(\Omega V') \otimes \mathbb{Q}$ if $H^*(V) \otimes \mathbb{Q}$ has more than one generator.

Proof. [32].   □

Conjecture. Let k be any field and V be a manifold such that $H^*(V, k)$ has more than one algebra generator, then the inclusion $V' \rightarrowtail V$ induces a surjection, $H_*(\Omega V'; k) \twoheadrightarrow H_*(\Omega V; k)$. (About the most recent results in this direction see [27]).

Chapter 7
Elliptic Spaces

In this chapter we shall put the theory of Sullivan models at work to prove non-trivial theorems in topology.

Let X be a 1-connected CW-complex with finitely many cells in each dimension. It is well-known that, for each i, its i-th homotopy group $\pi_i(X)$ is a finitely generated abelian group [52]).

The theorems we are about to state deal with the following question: Which influence do finiteness hypotheses on X have on $\pi_*(X)$? A rough answer is:

Principle: If X is small, then either $\pi_*(X)/(\text{torsion})$ is very small or very large.

Chapters 7 and 8 will give some precise answers to the above question. In this chapter our final goal is the following theorem.

0.1 Theorem. If $H_i(X;\mathbb{Q}) = 0$ for $i > n$ and if $\dim \pi_*(X) \otimes \mathbb{Q} < \infty$, then $\pi_i(X)$ is a finite abelian group for $i \geq 2n$.

We now will enter into a deeper study of Sullivan minimal models (cf. §1.2.1, Chapter 4). We begin with some purely algebraic matter, namely with formal dimension and related questions.

§1 Finiteness of the formal dimension

We denote by $(\Lambda V, d)$ such a model and we assume in the following that all minimal models satisfy $V^1 = 0$ and that V has finite type, i.e. the topological X is 1-connected and $H^i(X;\mathbb{Q})$ is finite dimensional for every i.

1.1 Definition. The *formal dimension* $fd(\Lambda V)$ of a commutative differential graded algebra $(\Lambda V, d)$ is equal to the maximum of all k such that $H^k(\Lambda V) \neq 0$. If no such k exists we write $fd(\Lambda V) = \infty$.

To begin with, we describe the important process which consists in killing variables in Sullivan models.

Suppose we are given a minimal Sullivan model $(\Lambda V, d)$; there always exists a $v \in V$ such that $dv = 0$. Let us choose a $\mathbb{Q}$-vector space W such that V decomposes as $V = \mathbb{Q}v \oplus W$ with $dv = 0$, and define $(\Lambda W, d) = \Lambda v \otimes \Lambda W/(v)$.

The algebraic structure is given by:

$$\Lambda V = \Lambda v \otimes \Lambda W = (\mathbb{Q} \oplus \Lambda^+ v) \otimes \Lambda W = \Lambda W \oplus \Lambda^+ v \otimes \Lambda W,$$

where the second summand is precisely the ideal $v.\Lambda V = (v)$, the ideal generated by v in ΛV. Because $dv = 0$, $v\Lambda V$ is d-stable $(d(v.\Lambda V) \subset v.d(\Lambda V))$. Then $\Lambda V/v.\Lambda V$ is a CDA; as a graded algebra it is isomorphic to ΛW. Using this isomorphism, we endow ΛW with a differential; let us denote this differential by $\bar{d}$. We say that $(\Lambda W, \bar{d})$ is the minimal Sullivan model obtained by killing the variable v. We now study the relation between the homologies of ΛW and ΛV.

1.2 Proposition. Suppose that ΛV decomposes as $\Lambda v \otimes \Lambda W$.

i) If $\deg v = 2n$ and $fd(\Lambda V) = k < \infty$, then $fd(\Lambda W) = k + 2n - 1$.

ii) If $\deg v = 2n + 1$ and $fd(\Lambda V) = k < \infty$, then

$$fd(\Lambda W) = k - (2n + 1) \quad \text{or} \quad fd(\Lambda W) = \infty.$$

Proof. We consider two cases.

1) $\deg v = 2n$. $\Lambda V = \Lambda[v]$, the polynomial algebra on one generator v. We have a short exact sequence

$$0 \longrightarrow \Lambda V \xrightarrow{v\wedge} \Lambda V \xrightarrow{\rho} \Lambda W \longrightarrow 0$$

and an associated long exact sequence (*Gysin sequence*)

$$\longrightarrow H^i(\Lambda V) \xrightarrow{[v]\wedge} H^{i+2n}(\Lambda V) \xrightarrow{H(\rho)} H^{i+2n}(\Lambda W)$$
$$\downarrow$$
$$H^{i+1}(\Lambda V) \longrightarrow$$

If $H^i(\Lambda V) = 0$ for $i > k$ and $H^k(\Lambda V) \neq 0$ for $i = k$, then $H^i(\Lambda W) = 0$ for $i > k + 2n - 1$ and $H^{k+2n-1}(\Lambda V) \neq 0$. This proves i) of Proposition 1.2.

2) $\deg v = 2n + 1$, $\Lambda v = \mathbb{Q} \oplus \mathbb{Q} v$ with $v^2 = 0$. We have a short exact sequence

$$0 \longrightarrow \Lambda W \xrightarrow{\Lambda v} \Lambda v \otimes \Lambda W \xrightarrow{\rho} \Lambda W \longrightarrow 0$$

Let us write $d(1 \otimes w) = 1 \otimes \bar{d}w \oplus v \otimes \theta w$. We thus define a linear map $\theta : \Lambda W \longrightarrow \Lambda W$ of degree $-2n$ which satisfies the following:

1.2.1 Lemma.

i) θ is a derivation of degree $-2n$ with the properties: $\theta \bar{d} = \bar{d}\theta$ and $\operatorname{Im}\theta \subset \Lambda^+ W$.

ii) Conversely, given θ as in i) we get a Sullivan model $(\Lambda V, d)$.

iii) $H(\theta) = \delta_*$ is the connecting morphism of the following exact sequence (*Wang sequence*)

$$\longrightarrow H^i(\Lambda W) \longrightarrow H^{i+2n+1}\Lambda V) \xrightarrow{H(\rho)} H^{i+2n+1}(\Lambda W)$$
$$H(\theta) \downarrow$$
$$H^{i+1}(\Lambda W) \longrightarrow$$

Proof. Easy and left as an exercise. The Wang sequence merely is the long exact cohomology sequence associated to

$$0 \longrightarrow \Lambda W \xrightarrow{\Lambda v} \Lambda v \otimes \Lambda W \xrightarrow{\rho} \Lambda W \longrightarrow 0.$$

□

Now, we calculate

$$d(v \otimes \Phi) = -v \otimes d\Phi.$$

So the morphism

$$\Lambda W \longrightarrow \mathbb{Q} v \otimes \Lambda W$$
$$\Phi \longmapsto v \otimes \Phi$$

is an isomorphism of degree $2n + 1$ of differential vector spaces. Looking at the Wang sequence above we conclude: If $H^i(\Lambda W) = 0$ for $i > N$ and $H^N(\Lambda W) \neq 0$, then $H^i(\Lambda V) = 0$ for $i > N + 2n + 1$ and $H^{N+2n+1}(\Lambda V) \neq 0$.

This proves that, if $fd(\Lambda V) = k$, then $fd(\Lambda W) > k - (2n + 1)$ implies $fd(\Lambda W) = \infty$.

□

Let us now look more closely at the case $fd(\Lambda V) = k < \infty$ and $\deg v = 2n+1$. Suppose $[z] \in H(\Lambda W)$ is a homology class with $\deg[z] > k - (2n + 1)$. Then

$$d(v \otimes z) = -v.(1 \otimes \bar{d}z + v \otimes \theta z) = 0$$

(recall that $v^2 = 0$) and $v \otimes z$ is of dimension $> k$.

Since ΛV has formal dimension n, $v \otimes z$ is a boundary and we can write

$$v \otimes z = d(v \otimes a + 1 \otimes z_1), \quad \text{for some } a \text{ and } z_1 \text{ in } \Lambda W$$
$$= -v \otimes \bar{d}a + 1 \otimes \bar{d}z_1 + v \otimes \theta z_1$$

and necessarily

$$\bar{d}z_1 = 0, \quad \theta z_1 = z + \bar{d}a.$$

So

$$H(\theta)[z_1] = [z] \quad \text{with} \quad \deg[z_1] = \deg[z] + 2n > k - (2n + 1) + 2n.$$

If there exists such a $[z]$ we get an infinite sequence

$$[z] \xleftarrow{H(\theta)} [z_1] \xleftarrow{H(\theta)} [z_2] \xleftarrow{H(\theta)} [z_3] \xleftarrow{H(\theta)}.$$

If the formal dimension of ΛW is infinite, we can choose $[z] = [z_0]$ to be a non-zero class. Thus we obtain an infinite sequence of non-zero cohomology classes $[z_i]_{i \in \mathbb{N}}$, with $\deg[z_i] = \deg[z_{i-1}] + 2n$.

On the other hand, recall we have defined $\theta : W \longrightarrow \Lambda^{\geq 1} W$. We can write $\theta = \theta_1 + \omega$ as a sum of two derivations

$$\theta_1 : \Lambda^q_* W \longrightarrow \Lambda^q_* W$$
$$\omega : \Lambda^q_* W \longrightarrow \Lambda^{>q}_* W.$$

We then can prove

1.3 Proposition. Assume $fd(\Lambda V, d) = k$. Then each of the assertions 1), 2), 3) implies the following one:

1) $fd\Lambda W = \infty$.

2)   There exists an infinite sequence of non-zero elements α_i of $H(\Lambda W)$ such that
$$\alpha_i = H(\theta)\alpha_{i+1}.$$

3)   There exists an infinite sequence of non-zero elements w_i of W such that
$$w_i = \theta_1 w_i.$$

4) $(\theta_1)^N|W \neq 0$ for all N.

Proof. The implication $(1) \Rightarrow (2)$ is already proved; $(3) \Rightarrow (4)$ is straightforward. We leave $(2) \Rightarrow (3)$ to the reader and prove $(2) \Rightarrow (4)$.

If α is cohomology class in $H(\Lambda W)$, let us define $e(\alpha) = \max r$, such that α has a representing element in $\Lambda^{\geq r} W$.

Then clearly
$$e(\alpha) \leq \deg \alpha.$$

Now by hypothesis we have a sequence of elements such that
$$\alpha \xleftarrow{H(\theta)} \alpha_1 \xleftarrow{H(\theta)} \alpha_2 \xleftarrow{H(\theta)} \alpha_3 \xleftarrow{H(\theta)} .$$

We first observe that $e(\alpha_i) \geq e(\alpha_{i+1})$. In fact, let r be maximal such that α_{i+1} is represented by $z_{i+1} \in \Lambda^{\geq r}W$. Then $\theta(z_{i+1}) \in \Lambda^{\geq r}W$ and represents α_i, therefore $e(\alpha_i) \geq r = e(\alpha_{i+1})$. We may thus assume that $e(\alpha_i) = r$ for all i.

Now every element $z \in \Lambda^{\geq r}W$ can be written uniquely in the form $z = u + \Phi$, for some $u \in \Lambda^r W$ and $\Phi \in \Lambda^{>r}W$. If we apply θ to z we obtain
$$\theta z = \underbrace{\theta_1 u}_{\in \Lambda^r W} + \underbrace{\omega(u)}_{\in \Lambda^{>r}W} + \underbrace{\theta\Phi}_{\in \Lambda^{>r}W} .$$

i.e. $\theta z = \theta_1 u + (\omega(u) + \theta(\Phi))$ is the decomposition of θz in $\Lambda^r W \oplus \Lambda^{>r}W$.

Therefore we have for all $j > 0$
$$\theta^j z - \theta_1^j u \in \Lambda^{>r}W.$$

Suppose $(\theta_1)^N|_W = 0$. Then
$$(\theta_1)^{Nr}(w_1 \wedge \cdots \wedge w_r) = \sum_{n_1+\ldots n_r = N_r} \pm(\theta_1)^{n_1} w_1 \wedge \cdots \wedge (\theta_r)^{n_r} w_r = 0.$$

Hence $(\theta_1)^{Nr} = 0$ on $\Lambda^r W$. Thus if $z = u + \Phi$ represents α_{Nr} with $u \in \Lambda^r W$, $\Phi \in \Lambda^{>r}$, we have
$$\theta^{Nr} z - (\theta_1)^{Nr}(u) \in \Lambda^{>r}W.$$

Now $(\theta_1)^{Nr}(u) = 0$. Thus $\theta^{Nr} z$ represents α and belongs to $\Lambda^{>r}W$; i.e. $e(\alpha) > r$ and this contradicts the hypothesis.   □

§2 Elliptic models

2.1 Definition. An *elliptic* Sullivan minimal model is a minimal model $(\Lambda V, d)$ such that

$$\dim V < \infty \quad \text{and} \quad \dim H(\Lambda V) < \infty.$$

For such models we can choose a finite basis $v_1, \ldots, v_r$ such that dv_i is a polynomial in the variables $v_1, \ldots, v_{i-1}$.

For each i we have the quotient model

$$(\Lambda(v_i, \ldots, v_r), \bar{d}) = (\Lambda V, d)/(v_1, \ldots, v_{i-1}).$$

2.2 Proposition. Suppose $(\Lambda V, d)$ is elliptic. Then

i)   For each i, $(\Lambda(v_i, \ldots, v_r), \bar{d})$ is elliptic.

ii) $fd(\Lambda V) = \displaystyle\sum_{|v_i|\,\text{odd}} \deg v_i - \sum_{|v_i|\,\text{even}} (\deg v_i - 1)$.

Proof. Let $W = (v_2, \ldots, v_r)$. If $\deg v_1$ is even, then by Proposition 1.2 ΛW is elliptic and $fd\Lambda W = fd(\Lambda V) + (\deg v_1 - 1)$.

If $\deg v_1$ is odd, let us consider the derivation $\theta : \Lambda W \longrightarrow \Lambda W$ as in 1.2 and its "linear part"

$$\theta_1 : \Lambda^q W \longrightarrow \Lambda^q W.$$

Since θ_1 decreases the degree, there exists an N such that one actually has $(\theta_1|_W)^N = 0$ $(N \leq r-1)$. By Proposition 1.3, this implies $H(\Lambda W) < \infty$; i.e. ΛW is elliptic and by Proposition 1.2 $fd(\Lambda W) = fd(\Lambda V) - \deg v_1$.

Repeating the process we get the announced formula.   □

Remark. Since $(\Lambda v_r, 0)$ is elliptic $|v_r|$ must be odd!   □

We now proceed to a characterization of elliptic models. Let P be the span of the odd generators $x_1, x_2, \ldots, x_k$ with $x_i = v_{2i+1}$. Let Q be the span of the even generators $y_1, y_2, \ldots, y_l$ with $y_i = v_{2i}$. We identify

$$\Lambda V = \Lambda Q \otimes \Lambda P = k[y_1, y_2, \ldots, y_l] \otimes E(x_1, x_2, \ldots, x_k)$$

$$= k[y_1, y_2, \ldots, y_l] \oplus \Lambda V \otimes P.$$

In particular, $dy_i \in \Lambda V \otimes P$ and we can write

$$dx_i = f_i(y) + \Omega_i \tag{$*$}$$

where f_i is a polynomial and $\Omega_j \in \Lambda V \otimes P$.

Define $n = \Sigma \deg x_i - \Sigma(\deg y_j - l)$.

2.3 Theorem. The following conditions are equivalent for a CDA $(\Lambda V, d)$ with $\dim V < \infty$:

i) $\dim H(\Lambda V) < \infty$; i.e. $(\Lambda V, d)$ is elliptic.

ii) $k[y_1, y_2, \ldots, y_l]/(f_1, \ldots, f_k)$ is finite dimensional.

iii) $H^n(\Lambda V) \neq 0$ and $H^i(\Lambda V) = 0$, for $n < i < 3n$.

Remark. iii) shows that ellipticity is decidable in finite time. We will derive iii) as a consequence of another theorem in the next chapter.

Proof of i) ⇔ ii).

i) ⇒ ii) It is enough to show that there exists an N, with

$$(y_m)^N \in (f_1, \ldots, f_k, y_1, \ldots, y_{m-1}), \quad \text{for all } m.$$

We first analyze the differential d: If $\Phi \in \Lambda V$ has odd degree, we write

$$\Phi = \sum_{i=1}^{k} g_i(y) \otimes x_i + \sum g_{i_1 i_2 i_3}(y) \otimes x_{i_1} x_{i_2} x_{i_3} + \ldots$$

Applying $(*)$ we see that

$$d\Phi = \sum_{1}^{k} g_i(y) f_i(y) + \Omega, \quad \text{for some } \Omega \in \Lambda V \otimes P.$$

Now kill the variables $x_1, \ldots, x_j, y_1, \ldots, y_{m-1}$ (the order of the variables is compatible with the degree). Then by Proposition 2.2 the quotient DGA:

$$(\Lambda(x_{j+1}, \ldots, x_k, y_m, \ldots, y_l), \bar{d})$$

is elliptic. Thus there exists an integer N such that the homology class of $(y_m)^N$ in $H((\Lambda(x_{j+1}, \ldots, x_k, y_m, \ldots, y_l), \bar{d})$ is zero. Thus

$$(y_m)^N = \bar{d}\Phi.$$

for some $\Phi \in \Lambda(x_{j+1}, \ldots, x_k, y_m, \ldots, y_l)$

By definition of $\bar{d}$, we can write $\bar{d}\Phi = d\Phi + \sum_{1}^{j} x_p \bar{\Phi}_p + \sum_{1}^{m-1} y_q \bar{\Psi}_q$ for some $\bar{\Phi}_q \bar{\Psi}_q$ in ΛV. Now let us decompose

$$\sum_{1}^{m-1} y_q \bar{\Psi}_q = \sum_{1}^{m-1} y_q h_q(y) + \Omega_1$$

with $\Omega_1 \in \Lambda V \otimes P$ and use the expression above for $d\Phi$ to get the formula

$$(y_m)^N = \sum_{1}^{k} g_i(y) f_i(y) + \sum_{1}^{m-1} y_q h_q(y) + \Omega + \Omega_1.$$

This formula implies that $\Omega + \Omega_1 = 0$ in $\Lambda V \otimes P$ and $(y_m)^N \in (f_1, \ldots, f_k, y_1, \ldots, y_{m-l})$, which is the first implication.

ii) ⇒ i). Let us define a new Sullivan model $(\Lambda V, \delta)$ by

$$\delta y_i = 0, \quad \delta x_i = f_i.$$

Then $\Lambda V = k[y_1, y_2, \dots, y_l] \otimes E(x_1, x_2, \dots, x_k)$ is finitely generated as a $k[y_1, y_2, \dots, y_l]$-module; moreover, δ is $k[y_1, y_2, \dots, y_l]$-linear and so $H(\Lambda V, \delta)$ is finitely generated as a $k[y_1, y_2, \dots, y_l]$-module.

But let m be the action of $k[y_1, y_2, \dots, y_l]$ on $H(\Lambda V, \delta)$. Then m factors by $\bar{m}$

$$k[y_1, y_2, \dots, y_l] \otimes H(\Lambda V) \xrightarrow{\;\;\;m\;\;\;} H(\Lambda V, \delta)$$
$$\searrow \qquad \nearrow \bar{m}$$
$$(k[y_1, y_2, \dots, y_l]/(f_1, \dots, f_k)) \otimes H(\Lambda V)$$

By ii) $k[y_1, y_2, \dots, y_l]/(f_1, \dots, f_k)$ is finite dimensional. Therefore so is $H(\Lambda V, \delta)$.

Finally, write

$$(\Lambda V)^r_{-s} = (\Lambda Q \otimes \Lambda^s P)^{r-s} \text{ for } 0 \le s \le k.$$

Then define $F^r(\Lambda V) = (\Lambda V)^{\ge r}_*$. One can easily check the following:
a) F^r is a filtration of differential algebras;
b) $(E_0, d_0) = (\Lambda V, \delta)$;
c) the induced spectral sequence is such that $E_{k+2} = E_\infty$.

The last point proves that the spectral sequence converges to $H(\Lambda V, d)$. Since $H(\Lambda V, \delta)$ is finite dimensional, so is $H(\Lambda V, d)$. □

Example. Let $\Lambda(x, y, z, a, u, v)$ be the free graded commutative algebra generated by x, y, z, a, u, v in respective degrees 3, 3, 3, 8, 13, 15. Define a differential by

$$dx = dy = dz = 0$$
$$da = xyz$$
$$du = xya$$
$$dv = a^2 + 2zu.$$

Here: $\delta x = \delta y = \delta z = \delta a = \delta u = 0$, $\delta v = a^2$. We obtain

$$H(\Lambda V, \delta) = \Lambda(x, y, z, u) \otimes H(\Lambda(a, v), \delta v = a^2)$$
$$= \Lambda(x, y, z, u) \otimes \Lambda a/a^2,$$

which is of dimension $32 < \infty$. Thus $H(\Lambda V, d)$ is finite dimensional and

$$fd(\Lambda V, d) = 3 + 3 + 3 - 7 + 13 + 15 = 30.$$

§3 Some equalities and inequalities

Suppose $(\Lambda V, d)$ is elliptic and generated by odd-degree generators $x_1, x_2, \ldots, x_k$, and even-degree generators $y_1, y_2, \ldots, y_l$ (we preserve all notations of the previous paragraph). Set

$$\deg y_i = 2a_i, \quad \deg x_j = 2b_j - 1 \quad \text{and} \quad n = fd(\Lambda V, d).$$

Recall from Proposition 2.2 ii):

Fact one. $n = \sum_{j=1}^{l}(2b_j - 1) - \sum_{i=1}^{k}(2a_i - 1).$

Next write $dx_j = f(y_1, y_2, \ldots, y_l) + \Lambda V \otimes P$ and recall that

$$\dim k[y_1, y_2, \ldots, y_l]/(f_1, \ldots, f_k) < \infty, \quad \deg f_j = 2b_j.$$

This means $k \geq l$, i.e.:

Fact two. The number of odd generators $\geq$ the number of even generators.

Now we renumber the indices such that

$$a_1 \geq \cdots \geq a_l \quad \text{and} \quad b_1 \geq \cdots \geq b_k$$

Define a map from $k[y_1, y_2, \ldots, y_l]/(f_1, \ldots, f_k)$ to
$k[y_1, y_2, \ldots, y_r]/(f_1(y_1, y_2, \ldots, y_r, 0, \ldots, 0), \ldots, f_k(y_1, y_2, \ldots, y_r, 0, \ldots, 0))$ by

$$y_i \longmapsto 0, \quad i > r.$$

At least r of the $f_j(y_1, y_2, \ldots, y_r, 0, \ldots, 0)$ are non-zero (because the last quotient algebra is finite, there are more equations than variables); say $f_{j_1}, \ldots, f_{j_r}$ out of $f_1, \ldots, f_k$. Thus $j_r \geq r$ and so

$$2b_r = \deg f_r \geq \deg f_{j_r} = \deg(y_1^{m_1}) \ldots (y_r^{m_r}) = \sum_i^r m_i 2a_i \geq 4a_r$$

and

$$b_r \geq 2a_r, \ 1 \leq r \leq l. \tag{$*$}$$

Now facts one and two combined yield

$$n \geq \sum_{j=1}^{l}(2b_j - 1) - \sum_{i=1}^{l}(2a_i - 1) = \sum_{j=1}^{l} 2(b_j - a_j).$$

Substituting the inequality $(*)$ we get:

$$n \geq \sum_{j=1}^{l} 2a_j.$$

That is

Fact three. $n \geq \sum_{i=1}^{l} \deg y_i$.

Finally, we obtain

$$n = \sum_{j=1}^{k} b_j + \sum_{j=1}^{k}(b_j - 1) - \sum_{i=1}^{l}(2a_i - 1)$$

$$\geq \sum_{j=1}^{k} b_j + \sum_{i=1}^{l}(b_i - 2a_i) \geq \sum_{j=1}^{k} b_j \quad \text{by inequality (∗)}.$$

Fact four. $2n - 1 \geq \sum_{j=1}^{k} \deg x_j$.

3.1 Corollary. If $(\Lambda V, d)$ is elliptic and has formal dimension n, then

$$V^q = 0, \quad q \geq 2n.$$

§4 Topological interpretation

4.1 Definition. A simply connected topological space X is *rationally elliptic* if

$$\dim H^*(X; \mathbb{Q}) < \infty \quad \text{and} \quad \pi_*(X) \otimes \mathbb{Q} < \infty.$$

Let us call $n = \max\{i; H^i(X; \mathbb{Q}) \neq 0\}$ the *formal dimension* of X (cf. §1.1).

We recall from Chapter 5 (especially §3) that the Sullivan minimal model of any rational 1-connected space X is a free CDA ΛV which satisfies
i) $H_*\Lambda V \cong H_*(X, \mathbb{Q})$.
ii) $\Lambda^+ V / \Lambda^+ V . \Lambda^+ V = \pi_*(X; \mathbb{Q})^{\#}$.
Now we can state:

4.2 Theorem. Suppose X is rationally elliptic of formal dimension n. Then
i) $\dim \pi_{\mathrm{odd}}(X) \otimes \mathbb{Q} \geq \dim \pi_{\mathrm{even}}(X) \otimes \mathbb{Q}$.
ii) If $\{x_j\}$ is a basis of $\pi_{\mathrm{odd}}(X) \otimes \mathbb{Q}$ and $\{y_i\}$ a basis of $\pi_{\mathrm{even}}(X) \otimes \mathbb{Q}$, then

$$n = \sum \deg x_j - \sum (\deg y_i - 1).$$

iii) $n \geq \sum \deg y_i$ and $2n - 1 \geq \sum \deg x_j$.
iv) $\pi_i(X) \otimes \mathbb{Q} = 0$, for $i \geq 2n$.

Proof. That is a mere transcription of results of §3.   □

4.3 Corollary. Theorem 0.1.   □

Chapter 8
Non Elliptic Finite C.W.-Complexes

In this chapter, we introduce the Lusternik-Schnirelmann category of a space. Then we show how this homotopy invariant is related to algebraic invariants of the Sullivan model of the space and its homotopy Lie algebra.

In contrast to the previous chapter, here we consider 1-connected spaces X such that $\dim \pi_*(X) \otimes \mathbb{Q} = \infty$.

Let us recall (cf. Chapter 2, §2.6.6) that on the graded group $\pi_*(X)$ there is defined the *Whitehead product*

$$[,] : \pi_i(X) \times \pi_j(X) \longrightarrow \pi_{i+j+1}(X).$$

which induces

$$[,]_{\text{free}} : \pi_i(X)/\text{torsion} \times \pi_j(X)/\text{torsion} \longrightarrow \pi_{i+j+1}(X)/\text{torsion}.$$

This product satisfies graded versions of the antisymmetry and Jacobi identities.

If $\alpha, \beta \in \pi_*(X)$ we shall write as usual $\operatorname{ad}(\alpha)(\beta) = [\alpha, \beta]$.

The following result expresses the fact that if X is a finite complex and if $\dim \pi_*(X) = \infty$, the Lie algebra structure on $\pi_*(\Omega X)$ is far from being nilpotent (the proof constitutes the last part of this chapter).

Let us call an element $\alpha \in \pi_{2i+1}(X) \otimes \mathbb{Q}$ "good" if there exists $\beta \in \pi_*(X) \otimes \mathbb{Q}$ with:

$$\forall k, \quad \operatorname{ad}^k(\alpha)(\beta) \neq 0.$$

0.1 Theorem. If $\dim X < \infty$, there exists an N such that, for every $i > N$, every non-zero $\alpha \in \pi_{2i+1}(X) \otimes \mathbb{Q}$ is good.

A related result is the following:

0.2 Theorem. If $H_i(X; \mathbb{Q}) = 0$ for $i > n$ and if $\dim \pi_*(X) \otimes \mathbb{Q} = \infty$, then in every interval $(j, j+n)$ of length n there is an integer k such that $\pi_k(X) \otimes \mathbb{Q} \neq 0$.

Proof. See [31]. □

Theorems 0.1 and 0.2 hold (and are actually proved!) under the more general hypothesis that X has finite Lusternik-Schnirelmann category. We introduce this notion in the first paragraph.

Thereafter we will come back to formal dimension and related questions for Sullivan models.

§1 Homotopy invariants of spaces

X will always denote a simply connected space such that each $H_i(X; \mathbb{Q})$ is finite dimensional.

One of the important invariants of a space was already mentioned in Chapter 7, namely the Lusternik-Schnirelmann category. We simply give the definition and refer to [39] for further use in rational homotopy theory.

1.1 Definition. The *Lusternik-Schnirelmann category* of X is the least integer m such that X can be covered by $m+1$ open sets, each of them contractible in X.

Exercise. Suppose X is a CW-complex of the form (the symbol $\cup$ represents the attachment of cells)

$$X = pt \cup \{e^{k_1,\alpha}\} \cup \{e^{k_2,\beta}\} \cup \cdots \cup \{e^{k_n,\gamma}\}.$$

Then cat $X \leq n$. In particular

$$\text{cat } X \leq \dim X - 1.$$

Before returning to algebra, we recall some well-known facts about the homotopy Lie algebra.

1.2 Homotopy and homology of the loop space. Recall from Chapter 2, §1 that, for the loop space ΩX, the composition of loops μ defines on $H_*(\Omega X; k)$ the structure of a connected graded algebra (as usual X is simply connected and $H_0(\Omega X; k) = k$); the product is defined by:

$$H_*(\Omega X; k) \otimes H_*(\Omega X; k) \xrightarrow{\cong} H_*(\Omega X \times \Omega X; k) \xrightarrow{H_*(\mu)} H_*(\Omega X; k).$$

Here are some basic results mentioned in Milnor and Moore [42].

1.2.1
i) (Cartan-Serre). The Hurewicz homomorphism

$$\pi_*(\Omega X) \longrightarrow H_*(\Omega X; \mathbb{Z})$$

becomes an isomorphism

$$\pi_*(\Omega X) \otimes \mathbb{Q} \xrightarrow{\cong} PH_*(\Omega X; \mathbb{Q})$$

where $PH_*(\Omega X; \mathbb{Q})$ are the primitives[4] of $H_*(\Omega X; \mathbb{Q})$.

---

(4) **Remarks on cocommutative Hopf algebras.**
Let H be a Hopf algebra with codiagonal

$$\Delta : H \longrightarrow H \otimes H, \ \Delta\alpha = \alpha \otimes 1 + \sum \alpha_i' \otimes \alpha_i'' + 1 \otimes \alpha.$$

α is primitive if $\Delta\alpha = \alpha \otimes 1 + 1 \otimes \alpha$. Moreover we have the commutator

$$[\alpha, \beta] = \alpha\beta + (-1)^{|\alpha|\cdot|\beta|}\beta\alpha \text{ in } H$$

which makes H into a graded Lie algebra, and $P_*(H)$ is a Lie subalgebra.

Thus $H_*(\Omega X; \mathbb{Q}) = U(\pi_*(\Omega X) \otimes \mathbb{Q})$.

ii)   The combined isomorphism

$$\pi_*(X) \otimes \mathbb{Q} \xrightarrow{\cong} \pi_{*-1}(\Omega X) \otimes \mathbb{Q} \xrightarrow{\cong} PH_{*-1}(\Omega X; \mathbb{Q})$$

sends the Whitehead product in $\pi_*(X) \otimes \mathbb{Q}$ to the commutator (*Samelson product*) in $PH_*(\Omega X; \mathbb{Q})$.

In the framework of cocommutative Hopf algebra, we recall the theorem of Milnor-Moore:

1.2.2   Let H be a cocommutative Hopf algebra over $\mathbb{Q}$. Then the inclusion $P(H) \longrightarrow H$ extends to an isomorphism of Hopf algebras

$$UP(H) \xrightarrow{\cong} H.$$

1.2.3   Let L_X be the *rational homotopy Lie algebra* of X: $L_k(X) = \pi_{k+1}(X) \otimes \mathbb{Q}$, where the Lie bracket is the Whitehead product. Combining 1.2.1 and 1.2.2, we obtain:

$$H_*(\Omega X; \mathbb{Q}) = UL_X.$$

§2   Sullivan models and the (algebraic) Lusternik-Schnirelmann category

2.1 Filtration of ΛV by word length. Recall (Chapter 4, §1.8.1)

$$\Lambda V = k \oplus V \oplus \Lambda^2 V \oplus \cdots \oplus \Lambda^s V \oplus \ldots$$

Define as usual $\Lambda^{\geq s} V$ by $\overset{i \geq s}{\oplus} \Lambda^i V$. The ideals $\Lambda^{\geq s} V$ are stable by d and so define a filtration leading to a spectral sequence given by

$$E_0^{p,*} = (\Lambda^p V, 0)$$
$$E_1^{p,*} = (\Lambda^p V, d_2) \Rightarrow H(\Lambda V, d)$$
$$E_2^{p,*} = \mathrm{Ext}_{UL}^p(k, k).$$

This suggests that there is a correspondence:

Homological properties of $UL \longleftrightarrow$ Homological properties of $(\Lambda V, d)$.

To gain more insight, however, we need to do better than to form a spectral sequence. Consider the projection of CDAs

$$(\Lambda V, d) \xrightarrow{\rho} (\Lambda V / \Lambda V^{>m}, d).$$

By axiom C_4 (Chapter 3, §1.2), we get a factorization

$$(\Lambda V, d) \xrightarrow{\quad j \quad} (\Lambda V \otimes \Lambda W, d)$$

with $H(\eta)$ an isomorphism. Moreover, we can choose η to be surjective (cf. [30]).

2.2 Algebraic definition of the Lusternik-Schnirelmann category. The *Lusternik-Schnirelman category* of $(\Lambda V, d)$ is the least m such that j admits a retraction r of CDAs

$$(\Lambda V, d) \xrightarrow{\quad j \quad} (\Lambda V \otimes \Lambda W, d) \xrightarrow{\quad r \quad} (\Lambda V, d)$$

with $r \circ j = id$.

The following theorem is a fundamental result of the theory of Sullivan models.

2.3 Theorem (Félix-Halperin). If X is simply connected with rationalization $X_{\mathbb{Q}}$ and minimal model $(\Lambda V, d)$ then

$$\mathrm{cat}(\Lambda V, d) = \mathrm{cat}(X_{\mathbb{Q}}) \leq \mathrm{cat}\, X.$$

Notation. We denote $\mathrm{cat}(X_{\mathbb{Q}})$ by $\mathrm{cat}_0(X)$.

Proof. Cf. [21]. □

Remark. One can also define another invariant, namely $M\, \mathrm{cat}_0(X)$ as the least m such that r exists as a map of ΛW-module.

We can now draw the connection to the following question of Ganea:

$$\text{Is } \mathrm{cat}(X \times S^n) \text{ equal to } \mathrm{cat}\, X + 1?$$

Jessup proved the (a priori) weaker result $M\, \mathrm{cat}_0(X_{\mathbb{Q}} \times S^n) = M\, \mathrm{cat}_0(X_{\mathbb{Q}}) + 1$ ([37]). The following theorem ([33]):

$$M\, \mathrm{cat}_0(X) = \mathrm{cat}_0(X)$$

proves that the answer to Ganea's conjecture is yes if X is rational.

Theorem 2.3 makes it possible to establish a much deeper relationship:

$$\text{Homological properties of } UL(X) \longleftrightarrow \text{Finiteness properties of } X.$$

2.4 Finiteness properties.

2.4.1 Proposition. For any differential Lie algebra L, if $\operatorname{Ext}_{UL}^i(k,k) = 0$ then

$$\operatorname{Ext}_{UL}^j(k,k) = 0 \text{ for } j \geq i.$$

Proof. See [38], appendix.   □

We now define:

2.4.2 Definition.
1) The *global dimension* of L: $gl \dim L = \sup\{m; \operatorname{Ext}_{UL}^m(k,k) \neq 0\}$.
2) The *depth* of L: $\operatorname{depth} L = \inf\{m; \operatorname{Ext}_{UL}^m(\mathbb{Q}, UL) \neq 0\}$;
    $\operatorname{depth} L = \infty$ if $\operatorname{Ext}_{UL}^m(k,k) = 0$ for any m.

2.4.3 Theorem (Félix-Halperin-Jakobson-Löfwall-Thomas). If X is simply connected and each $H_i(X; \mathbb{Q})$ is finite dimensional, then

$$\operatorname{depth} L_X \leq \operatorname{cat} X.$$

Proof. Cf. [22].   □

In the following, we will state and prove much more refined versions of Theorem 0.1.

§3   Lie algebras of finite depth

3.1 Definition. Let L be a graded Lie algebra. Then α is an *Engel element* if for all $\beta \in L$ there exists an $n(\beta)$ such that

$$\operatorname{ad}(\alpha)^{n(\beta)}\beta = [\alpha, [\alpha, [\alpha, \ldots [\alpha, \beta] \ldots]]] = 0.$$

In other words an Engel element is an element which is not good in the sense of the introduction.

We denote by E_r the linear span of the Engel elements of degree r.

3.2 Theorem (Félix-Halperin-Thomas). Let L be a graded Lie algebra. Then

$$\sum_q \dim E_{2q} \leq \operatorname{depth} L.$$

Proof. Cf. [24].   □

3.2.1 Corollary. Any basis $\{\alpha_i\}$ of L_{even} contains at most $d = \operatorname{depth} L$ Engel elements; i.e. except for $\alpha_{i_1}, \ldots, \alpha_{i_d}$, for each α_j there exists a β_j such that

$$(\operatorname{ad} \alpha_j)^n \beta_j \neq 0 \text{ for all } n.$$

□

3.3 Proposition.
i)   Let I be an ideal in a graded Lie algebra L. Then

$$\text{depth } I \le \text{depth } L.$$

ii)   An infinite dimensional abelian Lie algebra has infinite depth.

Proof. i) Follows directly from the Hochschild-Serre spectral sequence [35] whose E_2 term is

$$E_2^{p,q} = \text{Ext}_{U(L/I)}^p(\mathbb{Q}, \text{Ext}_{UL}^q(Q, UL))$$

and which converges to

$$\text{Ext}_{UL}^{p+q}(\mathbb{Q}, UL).$$

As UL is free as a UI-module, $\text{Ext}_{UI}^q(\mathbb{Q}, UL) = 0$ for $q < $ depth I. Then the same holds for $\text{Ext}_{UL}^q(\mathbb{Q}, UL)$.

ii) We first study the case where L is generated by only even elements. Suppose $L = \mathbb{Q}.x, |x|$ even; then the enveloping algebra is a polynomial algebra on the generator $x : UL = \mathbb{Q}[x]$. The resolution of $\mathbb{Q}$ as a $\mathbb{Q}[x]$-module:

$$0 \longrightarrow \mathbb{Q}[x].\alpha \overset{\partial}{\longrightarrow} \mathbb{Q}[x] \overset{\varepsilon}{\longrightarrow} \mathbb{Q} \longrightarrow 0, \partial(\alpha) = x$$

shows that

$$\text{Ext}_{UL}^q(\mathbb{Q}, UL)) = \begin{cases} 0 & \text{if } q \ne 1 \\ \mathbb{Q} & \text{if } q = 1 \end{cases},$$

generated by the application

$$\mathbb{Q}[x].\alpha \longrightarrow \mathbb{Q}[x]$$
$$\alpha \mapsto 1$$

If $L = \mathbb{Q}.x, |x|$ odd, the enveloping algebra is an exterior algebra on the generator $x: UL = E(x)$. The resolution of $\mathbb{Q}$ as an $E(x)$-module:

$$E(x).\alpha_p \overset{\partial}{\longrightarrow} \dots E(x).\alpha_2 \overset{\partial}{\longrightarrow} E(x).\alpha_1 \overset{\partial}{\longrightarrow} E(x).\alpha_0 \overset{\varepsilon}{\longrightarrow} \mathbb{Q} \longrightarrow 0,$$
$$\partial(\alpha_p) = x\alpha_{p-1}, \quad \varepsilon(\alpha_0) = 1$$

shows that

$$\text{Ext}_{UL}^q(\mathbb{Q}, UL)) = \begin{cases} 0 & \text{if } q \ne 0 \\ \mathbb{Q} & \text{if } q = 0 \end{cases},$$

generated by the application

$$E(x).\alpha_0 \longrightarrow E(x)$$
$$\alpha_0 \mapsto 1$$

By the identity

$$\text{Ext}_{UL \otimes UL'}(\mathbb{Q}, UL \otimes UL') = \text{Ext}_{UL}(\mathbb{Q}, UL) \otimes \text{Ext}_{UL'}(\mathbb{Q}, UL') \qquad (*)$$

only the case of an infinite abelian Lie algebra generated in odd degrees still needs a proof.

Consider such a Lie algebra L. Suppose we have proved $\text{Ext}_{UL}^0(\mathbb{Q}, UL) = 0$. Then depth $L \geq 1$. Now for every integer n we can choose infinite dimensional Lie subalgebras L_i, $i \leq n$, such that $L = \bigoplus_{i=1}^{n} L_i$; then $UL = \bigoplus_{i=1}^{n} L_i$ and $(*)$ shows that depth $L \geq n$. This holds for every n and so depth $L \geq \infty$.

It remains now to prove depth $L \geq 1$ for an infinite dimensional Lie algebra generated in odd degrees. The identification $\text{Ext}_{UL}^0(\mathbb{Q}, UL) = \{\alpha \in UL; \forall \beta \in L, \beta.\alpha = 0\}$ proves that $\text{Ext}_{UL}^0(\mathbb{Q}, UL) = 0$.   □

Let us return to the case where depth L is finite.

3.4 Theorem. Let L be a graded Lie algebra such that

$$\dim L = \infty \text{ and depth } L = d < \infty.$$

Then
i)   Any solvable[5] Lie ideal $I \subset L$ satisfies $\dim I_{\text{even}} \leq d$ and I is finite dimensional.
ii)   The sum of the solvable ideals of L is finite dimensional.
iii) L_{even} is infinite dimensional.

Proof. i) First suppose that I is abelian. Then every element of I is Engel and so by Theorem 3.2 and Proposition 3.3

$$\dim I_{\text{even}} \leq \text{ depth } I \leq \text{ depth } L = d.$$

Thus, for N big enough, $I_{\geq N}$ is concentrated in odd degrees and therefore abelian. Now, by Proposition 3.3 the depth of an infinite dimensional abelian ideal is infinite and so $I_{\geq N}$ is finite dimensional.

Finally, by induction, $I' = [I, I]$ is finite dimensional and so $I_{\geq N}$ is abelian, for N big enough. By the arguments above I is finite dimensional; thus every element of I is Engel and $\dim I_{\text{even}} \leq d$.
ii) First let us call the sum of the solvable ideals of a Lie algebra L the *radical* of L and denote it by rad L.

By i), $(\text{rad } L)_{\text{even}}$ is finite dimensional; thus for N big enough $(\text{rad } L)_{\geq N}$ possesses only elements of odd degrees and so is abelian. By Proposition 3.3 $(\text{rad } L)_{\geq N}$ is finite dimensional, which ends the proof of ii).
iii) Suppose L_{even} is of finite dimension N. Then $L_{\geq N}$ is an abelian ideal of L. By hypothesis depth L is finite and by Proposition 3.3 i) so is depth $L_{\geq N}$. By 3.3 ii) $L_{\geq N}$ is finite dimensional; a contradiction to the hypothesis $\dim L = \infty$.   □

---

(5) A Lie algebra L is *solvable* if the following sequence of ideals $I_0 = L$, $I_1 = [L, L]$, $I_2 = [I_1, I_1]$, stops at $I_j = 0$ for some integer j.

3.5 Theorem. Let X be simply connected with each $H_i(X; \mathbb{Q})$ finite. Assume

$$\dim \pi_*(X) \otimes \mathbb{Q} = \infty \text{ and } \operatorname{cat} X = m < \infty.$$

Then
i) The sum of the solvable ideals of $L(X)$ is a finite dimensional ideal R with

$$\dim R_{\text{even}} \leq \operatorname{cat} X.$$

ii) $L_{\text{even}}(X) = \infty$.

Proof. Simply translate Theorem 3.4 into the language of topology and use Theorem 2.4.3. □

§4 The mapping theorem

We return to minimal models, leaving aside for the moment our study of the homotopy Lie algebra.

4.1 Theorem. Let $\varphi : (\Lambda V, d) \longrightarrow (\Lambda W, d)$ be a surjection of minimal Sullivan models. Then

$$\operatorname{cat}(\Lambda V, d) \geq \operatorname{cat}(\Lambda W, d).$$

Proof. Cf. [21] and [25]. □

4.1.1 Corollary. Let $f : X \longrightarrow Y$ be a continuous map of topological spaces (simply connected) such that $\pi_*(f) \otimes \mathbb{Q}$ is injective. Then

$$\operatorname{cat}_0(X) \leq \operatorname{cat}_0(Y).$$

Proof. Just remember the connection between the Sullivan model and the homotopy Lie algebra (Chapter 5, §3); we also note that, if $(\Lambda V, d)$ (resp. $(\Lambda W, d)$) denotes the Sullivan model of X (resp. Y), $\pi_*(f) \otimes \mathbb{Q}$ injective is equivalent to $V \longrightarrow W$ surjective which is also equivalent to $\Lambda V \longrightarrow \Lambda W$ surjective. □

This theorem is the principal ingredient in the following

4.2 Theorem. Let $(\Lambda V, d)$ be a minimal model such that

$$\dim V = \infty \text{ and } \operatorname{cat}(\Lambda V, d) < \infty.$$

Set $n_k = \sum_{j \leq k} \dim V^j$. Then there exists a constant $C > 1$ and an integer N such that $n_r \geq C^r$ for $r > N$ (this property is referred to as the *exponential growth* of V).

Proof. See [23] which gives a more precise result. □

4.2.1 Corollary. Suppose X is simply connected and has finite category. Then either

$$\dim \pi_*(X) \otimes \mathbb{Q} \leq 2 \operatorname{cat}(X)$$

or else

$$\left[\sum_{i \leq r} \dim \pi_i(X) \otimes \mathbb{Q} \right] \geq C^r.$$

Proof. In the first case the space is elliptic and we know (Chapter 7, §4.2) that

$$\dim \pi_{\text{odd}} \geq \dim \pi_{\text{even}} \quad \text{and}$$
$$\dim \pi_{\text{odd}} = \dim L_{\text{even}} = \dim E_{\text{even}} \leq \operatorname{cat} X.$$

The second case is the topological formulation of the preceding statement.   □

A refined version of Theorem 4.2 leads to the following result:

4.3 Theorem. Let $X_{\langle n \rangle} \longrightarrow X \longrightarrow X_n$ be the n-th Postnikov fibration in the Postnikov tower (i.e. the homotopy decomposition; cf. Chapter 2). Then, if $H_i(X; \mathbb{Q}) = 0$ for $i > N$, then there exists a k such that:

$$H^*(X_{\langle n \rangle}; \mathbb{Q}) < \infty \quad \text{if } n < k,$$
$$= \infty \quad \text{if } n \geq k.$$

Proof. See [20], Chapter 6.   □

§5 Proof of Theorem 0.1.

We end this chapter with a sketch of the proof of Theorem 0.1. Consider again a minimal model $(\Lambda V, d)$ such that

$$\operatorname{cat}(\Lambda V, d) = m < \infty \quad \text{and} \quad \dim V = \infty.$$

Again we choose a basis

$$v_1, v_2, v_3, \dots \quad \text{with} \quad dv_i = g_i(v_1, \dots, v_{i-1}).$$

By killing $v_1, \dots, v_r$ we get a quotient model $\Lambda(V(r), \bar{d}) = \Lambda(v_r, \dots)$.

Example. $(\Lambda V, d) = (\Lambda v \otimes \Lambda W,)$, $\deg v = 2q + 1$.

Recall (Chapter 7, §1) that for $w \in W$ we have
$$d(1 \otimes w) = 1 \otimes \bar{d}w + v \otimes \theta w \ldots,$$
and
$$\theta = \theta_1 + \omega, \quad \text{where} \quad \begin{cases} \theta_1 : W \longrightarrow W \\ \omega : W \longrightarrow \Lambda^{\geq 2} W \end{cases}$$

Thus
$$d_2(1 \otimes w) = 1 \otimes \bar{d}_2 w + v \otimes \theta_1 w.$$

Now let $\alpha \in V^{*-1}$ be defined by $\alpha(v) = 1$, $\alpha(W) = 0$ and let $\beta \in W^*$ satisfy $\beta(v) = 0$. Then $\deg \alpha = 2q$ is even and
$$\langle [\alpha, \beta], v \rangle = \langle \alpha \otimes \beta, dv \rangle = 0 \quad \text{because} \quad dv = 0;$$
$$\langle [\alpha, \beta], w \rangle = \langle \alpha \otimes \beta, 1 \otimes \bar{d}_2 w + 1 \otimes \theta_1 w \rangle = \beta(\theta_1 w).$$

i.e.
$$\langle (\mathrm{ad}\,\alpha)\beta, w \rangle = \langle \beta(\theta_1 w) \rangle.$$

This shows that θ_1 is dual to $\mathrm{ad}\,\alpha$. In the preceding chapter we observed that if $H(\Lambda V) < \infty$ and $H(\Lambda W) = \infty$ then we had
$$W \xleftarrow{\theta_1} W_1 \xleftarrow{\theta_1} W_2 \xleftarrow{\theta_1} W_3 \xleftarrow{\theta_1}.$$
Choose β such that $\langle \beta, w \rangle = 1$ such that $(\theta_1)^k = w_k$. Then
$$\langle (\mathrm{ad}\,\alpha)^k \beta, w_k \rangle = \langle \beta, (\theta_1)^k w_k \rangle = \langle \beta, w \rangle = 1.$$
Thus $(\mathrm{ad}\,\alpha)^k \beta \neq 0$ for all k.

5.1 Proposition. If $\dim H(\Lambda V) < \infty$ and $\dim H(\Lambda W) = \infty$ there exist α and β such that
$$(\mathrm{ad}\,\alpha)^k \beta \neq 0 \text{ for all } k.$$

The preceding assertion leads to the following theorem in topology.

5.2 Theorem. Let X be a simply connected space such that
$$\dim H^*(X; \mathbb{Q}) < \infty \quad \text{and} \quad \dim L_*(X) = \dim \pi_*(X) \otimes \mathbb{Q} = \infty.$$
Then $\exists \alpha \in L_{2q}(X), \beta \in L_*(X)$ such that
$$(\mathrm{ad}\,\alpha)^k \beta \neq 0 \text{ for all } k.$$

Indications. Write the minimal model of X in the form
$$(\Lambda(v_1, v_2, v_3, \ldots), d).$$

Show that if you kill enough variables, the cohomology eventually becomes infinite. Thus get $\Lambda(v_r, v_{r+1}, \ldots) = \Lambda(v_r \oplus W)$ such that
$$H(\Lambda(v_r \oplus W)) < \infty \quad \text{and} \quad H(\Lambda W) = \infty.$$

By the preceding chapter v_r is odd, apply above example. For a complete proof see [24]. □

5.2.1 Corollary. $L_*(X)$ is not abelian. □

Chapter 9
Towards Integral Algebraic Models of Homotopy Types

In the preceding chapters, we essentially considered algebraic models of rational spaces. We would now like to seek an algebraic integral "model", especially for polyhedra of lower dimension.

§1 Introduction and general problem

From the beginning of these lectures, we have been concerned more or less with the homotopy classification problem. The first tools available are of course the homology and homotopy groups. To gain a grip on the topological structure of the spaces we have considered standard decompositions (by fibrations: Postnikov towers, or by cofibrations: CW-complexes) and tried to mimic them in certain algebraic categories. Because of the richness of the homotopy groups of spheres, we have only achieved this for rational spaces.

The main purpose of this book is to provide an introduction to rational homotopy theory. Indeed in this case we obtained theories of algebraic models which describe completely the rational homotopy category.

Our wish would be to extend some results in rational homotopy to integral homotopy. Then in general an algebraic model of homotopy type *should provide* a good procedure to compute the set $M(H_*)$ of all homotopy types of X with $H_*(X) \cong H_*$.

Let us remark that we have already (cf. Chapter 2, §3) determined the homotopy types of all spaces X with $H_*(X) = \mathbb{Z}$ if $* = n$ or $m+1$ ($n \le m$) and 0 otherwise ($X = S^n \cup_f e^{m+1}$ with $f \in \pi_m S^n$).

At this point we can summarize the various cases to be examined (a priori!) in order to model homotopy types.

	$\pi_1 \ne 0$	$\pi_1 = 0$	stable
$\mathbb{Z}$	Total (to hard in full detail) restrict to low dimension where $\pi_* S^*$ is known	Simply connected *unstable* homotopy theory	Stable integral homotopy theory
$\mathbb{Q}$	? Not clear	Rational homotopy theory	Trivial!

The first attempt to classify homotopy types over $\mathbb{Z}$ was achieved by J.H.C. Whitehead [51]. In his paper he gave a complete algebraic classification of 1-connected 4-dimensional CW-complexes. Baues [9] extended the classification to the non-connected case.

For any CW complex X let us consider the long exact sequences of homotopy associated to the pairs (X^n, X^{n-1})

$$\to \pi_m(X^n, X^{n-1}) \xrightarrow{\partial_m^n} \pi_{m-1}(X^{n-1}) \xrightarrow{i_{m-1}^n} \pi_{m-1}(X^n) \xrightarrow{j_{m-1}^n} \pi_{m-1}(X^n, X^{n-1}) \to .$$

In his paper J.H.C. Whitehead noticed that the following sequence

$$\pi_3(X^3, X^2) \xrightarrow{\ b\ } \pi_2(X^2, X^1) \xrightarrow{\ \partial\ } \pi_1 X^1 \quad \text{where} \quad b = j_2^2 \circ \partial_3^3 \quad \text{and} \quad \partial = \partial_2^2$$

describes the 3-dimensional complexes, but that

$$\pi_4(X^4, X^3) \longrightarrow \pi_3(X^3, X^2) \longrightarrow \pi_2(X^2, X^1) \longrightarrow \pi_1 X^1$$

does not detect the Hopf map, i.e. does not distinguish $\mathbb{C}P(2)$ from $S^2 \vee S^4$.

So we want to give a short description of Whitehead's "algebraic model" for the 4-dimensional polyhedra.

§2 Algebraic description of the integral homotopy types in dimension 4

2.1 The Γ-groups and the Whitehead exact sequence. Let X be a simply connected CW-complex with basepoint and let $SP^\infty(X)$ be the infinite symmetric product of X (cf. Chapter 2, §1.5.4). By the result of Dold-Thom (cf. Chapter 2, §1.5.5) we have the natural equation $\pi_n SP^\infty(X) = H_n(X)$. Moreover the canonical inclusion $j : X \rightarrowtail SP^\infty(X)$ induces the Hurewicz homomorphism h. Let us denote by ΓX the homotopy fibre of j. Therefore the homotopy fibration sequence of the fibration

$$\Gamma X \xrightarrow{\ i\ } X \xrightarrow{\ j\ } SP^\infty(X)$$

is (in low degrees)

$$H_4 \xrightarrow{\ b\ } \Gamma_3 \xrightarrow{\ i_*\ } \pi_3 \xrightarrow{\ h\ } H_3 \longrightarrow 0 = \Gamma_2 \xrightarrow{\ i_*\ } \pi_2 \xrightarrow{\ h\ } H_2.$$

2.1.1 Definition. The n-th Γ-*group* of X is defined by $\Gamma_n(X) = \pi_n(\Gamma X)$.

Let X^n be the n-skeleton of X; one can prove:

2.1.2 Proposition. $\Gamma_n(X) = \mathrm{Im}(\pi_n X_*^{n-1} \longrightarrow \pi_n X_*^n)$.

Proof. Cf. [51].   □

Since X is simply connected the Hurewicz homomorphism is an isomorphism in degree 2:

$$\pi_2 \xrightarrow[h]{\ \cong\ } H_2 \quad \text{and} \quad \Gamma_2 = 0.$$

Γ_3 was computed by Whitehead. He defined for that purpose the notion of universal quadratic functor.

2.2.1 Definition. A map $f : A \longrightarrow B$ between abelian groups is *quadratic* if $f(a) = f(-a)$ and if $f(a + b) - f(a) - f(b)$ is bilinear, for $a, b \in A$. There is a universal quadratic map

$$\gamma : A \longrightarrow \Gamma A$$

such that for each quadratic map f there is a unique homomorphism $\bar{f} : \Gamma A \longrightarrow B$ between abelian groups with $\bar{f} \circ \gamma = f$. This defines the functor Γ on abelian groups. The functor **Quadr** is representable, i.e.

$$\mathbf{Quadr}(A, B) = \mathbf{Ab}(\Gamma A, B).$$

Let A be an abelian group; J.H.C. Whitehead noticed that

$$\Gamma A = H^4(K(A, 2)) \, (= \pi_3(M(A, 2))$$

and therefore

$$\Gamma_3 = \Gamma(\pi_2(X)).$$

The Postnikov invariant k_3 is an element of

$$H^4(K(\pi_2(X)) = \pi_3(M(H_2, 2)) = \Gamma_3 = \Gamma(\pi_2(X)).$$

Moreover, from the fibre sequence above we can extract the following short exact sequence:

$$0 \longrightarrow \operatorname{coker} b \rightarrowtail \pi_3(X) \longrightarrow H_3(X) \longrightarrow 0$$

This shows that $\pi_3(X)$ represents an element of $\mathrm{Ext}(H_3, \operatorname{coker} b)$; let us denote it by $\{\pi_3(X)\}$.

Now look at the following diagram, where the first line is the universal coefficient theorem for homotopy groups with coefficients:

$$\longrightarrow \mathrm{Ext}(H_3, \pi_3(M(H_2, 2))) \overset{\Delta}{\rightarrowtail} [M(H_3, 2), M(H_2, 2)] \overset{\mu}{\longrightarrow\!\!\!\rightarrow} \mathrm{Hom}(H_3, H_2)$$

$$\downarrow$$

$$\mathrm{Ext}(H_3, \operatorname{coker} b)$$

The invariant k_3' of the homology decomposition is an element of $[M(H_3, 1), M(H_2, 2)]$ which goes to zero in $\mathrm{Hom}(H_3, H_2)$ and therefore comes from an element of $\mathrm{Ext}(H_3, \pi_3(M(H_2, 2)))$.

Conversely, any choice of $k' \in \mathrm{Ext}(H_3, \Gamma(\pi_2))$ gives a homology decomposition of a 4-dimensional complex. Actually, Whitehead exact sequences and the homotopy type of polyhedra are strongly related and Theorem 2.2.4, proved by J.H.C. Whitehead, states the precise result.

Let H denote a graded abelian group $H = \{H_n\}$ with $H_n = 0$ for $n < 0$ and $n > 4$ and such that H_4 is free.

Consider now the set of pairs $(b_4, \beta_4)^{(6)}$ defined with

$$b_4 \in \mathrm{Hom}(H_4, \Gamma(H_2))$$
$$\beta_4 \in \mathrm{Ext}(H_3, \mathrm{coker}\ b_4).$$

2.2.2 Definition. We say that two pairs (b_4, β_4) and (b_4', β_4') corresponding to the homology H are isomorphic if and only if there exists an automorphism φ : $H_* \cong H_*$ such that

$$\Gamma(\varphi_2)_* b_4 = \varphi_4 b_4'$$
$$\varphi_{4*} \beta_4 = \varphi_3^* \beta_4'.$$

On the other hand, given a 1-connected four-dimensional CW complex X, we associate the exact sequence of Whitehead

$$(E) : H_4 \xrightarrow{b} \Gamma_3 \xrightarrow{i_*} \pi_3 \xrightarrow{h} H_3 \longrightarrow 0.$$

2.2.3 Definition. We say that a triple $(f_4, f_3, f_2) : H_i \longrightarrow H_i'$ is a morphism of two sequences (E) and (E') if there exists a φ fitting into the commutative ladder

$$
\begin{array}{ccccccccc}
(E) : H_4 & \xrightarrow{\ b\ } & \Gamma_3 = \Gamma(H_2) & \xrightarrow{\ i_*\ } & \pi_3 & \xrightarrow{\ h\ } & H_3 & \longrightarrow & 0 \\
f_4 \downarrow \cong & & \Gamma(f_2) \downarrow \cong & & \varphi \downarrow \cong & & f_3 \downarrow \cong & & \\
(E) : H_4' & \xrightarrow{\ b'\ } & \Gamma_3' = \Gamma(H_2') & \xrightarrow{\ i_*'\ } & \pi_3' & \xrightarrow{\ h'\ } & H_3' & \longrightarrow & 0
\end{array}
$$

Such a morphism is an isomorphism of Whitehead sequences if each vertical arrow is an isomorphism.

The main theorem of J.H.C. Whitehead [51] now reads:

2.2.4 Theorem. The homotopy types of four dimensional complexes with homology H are in one-to-one correspondance with equivalence classes of pairs (b_4, β_4) defined in 2.2.2 where the algebra H moreover is supposed to be 1-connected $(H_0 = \mathbb{Z}$ and $H_1 = 0)$.

---

(6) We use the index "4" in order to agree with the general notation of §3.

Proof. See [51]. Whitehead proved that the homotopy types of 1-connected four-dimensional CW complexes are equivalent to isomorphism classes of Whitehead sequences as defined in 2.2.3. Here we just describe the correspondance between homotopy types of 1-connected four-dimensional CW complexes and pairs (b_4, β_4).

To a complex X, associate the pair (b_4^X, β_4^X), where b_4^X is the map $b \colon H_4 \longrightarrow \Gamma_3$ of the J.H.C. Whitehead sequence and $\beta_4^X = \{\pi_3(X)\}$ is the class of the extension

$$0 \longrightarrow \operatorname{Coker} b_r \rightarrowtail \pi_3(X) \longrightarrow H_3(X) \longrightarrow 0.$$

$\square$

§3   Algebraic description of the integral homotopy types in dimension N

We can try to extend the above determination of simply connected four-dimensional CW-complexes to the general case of CW-complexes of dimension N.

3.1 Γ-groups with coefficients. Recall that in Chapter 2 (§1.8) we defined the n-th homotopy group of a space X with coefficients in A by $\pi_n(A; X) = [M(A, n), X]$.

3.1.1 Definition. If A is an abelian group, the n-th Γ-*group of X with coefficients* in A is defined by $\pi_n(A, \Gamma X)$ and denoted by $\Gamma_n(A; X)$.

3.2 The general theorems. As in the preceding case we start from the exact sequence of J.H.C. Whitehead

$$\longrightarrow H_{n+1} \overset{b_{n+1}}{\longrightarrow} \Gamma_n \overset{i_n}{\longrightarrow} \pi_n \overset{h_n}{\longrightarrow} H_n \longrightarrow \ldots H_3 \longrightarrow 0.$$

From this we extract the following short exact sequence

$$\operatorname{coker} b_{n+1} \overset{i_n}{\rightarrowtail} \pi_n(X) \overset{h_n}{\twoheadrightarrow} \ker b_n.$$

This extension of abelian groups defines an element $\{\pi_n(X)\}$ in $\operatorname{Ext}(\ker b_n, \operatorname{coker} b_{n+1})$.

Let us denote $H_n(X)$ by H and consider the inclusion $i \colon \ker b_n^X \subset H_n X$; i induces

$$i^{\#} = \operatorname{Ext}(i, \operatorname{coker} b_{n+1}) \colon \operatorname{Ext}(H, \operatorname{coker} b_{n+1}) \longrightarrow \operatorname{Ext}(\ker b_n, \operatorname{coker} b_{n+1}).$$

We can construct the following diagram

$$
\begin{array}{ccccccc}
\operatorname{Ext}(H, H_{n+1}X) & \overset{\Delta}{\rightarrowtail} & \pi_n(H, SP^\infty(X)) & \overset{\mu}{\longrightarrow} & \operatorname{Hom}(H, H_n X) & & \ni 1 \\
\downarrow{\scriptstyle \operatorname{Ext}(b, b_{n+1})} & & \downarrow{\scriptstyle \partial} & & \downarrow & & \big\uparrow \\
\operatorname{Ext}(H, \Gamma_n X) & \overset{\Delta}{\longrightarrow} & \pi_{n-1}(H; \Gamma X) & \overset{\mu}{\longrightarrow} & \operatorname{Hom}(H, \Gamma_{n-1} X) & & \ni b_n^X \\
\downarrow & {\scriptstyle \text{p.o.}} & \downarrow & {\scriptstyle \mu} & \nearrow & & \\
\operatorname{Ext}(H, \operatorname{coker} b_{n+1}) & \overset{\Delta}{\longrightarrow} & \Gamma_{n-1}^b(H; X) & & \ni \beta_n^X & &
\end{array}
$$

Here the left column is induced by the composite

$$H_{n+1}X \xrightarrow{b_{n+1}} \Gamma_n X \twoheadrightarrow \Gamma_n X / \operatorname{Im} b_{n+1}.$$

$\partial : \pi_n(H; SP^\infty(X)) \longrightarrow \pi_{n-1}(H; \Gamma X)$ is the connecting morphism of the homotopy long exact sequence of the fibration

$$\Gamma X \xrightarrow{i} X \xrightarrow{j} SP^\infty(X).$$

The right column is induced by $b_{n-1} : H_n X \longrightarrow \Gamma_{n-1} X$.

The top and middle lines are the universal coefficient sequences for homotopy groups with coefficients [41].

$\Gamma^b_{n-1}(H; X)$ is defined as the push-out

$$
\begin{array}{ccc}
\operatorname{Ext}(H, \Gamma_n X) & \xrightarrow{\quad \Delta \quad} & \pi_{n-1}(H; \Gamma X) \\
\downarrow & \text{p.o.} & \downarrow \\
\operatorname{Ext}(H, \ \operatorname{coker} b_{n+1}) & \xrightarrow{\quad \Delta \quad} & \Gamma^b_{n-1}(H; X)
\end{array}
$$

3.2.1 Theorem. To each 1-connected CW-complex X there is canonically associated a sequence of elements

$$\beta = (\beta_4^X, \beta_5^X, \dots) \quad \text{with} \quad \beta_{n+1}^X \in \Gamma^b_{n-1}(H_n X; X)$$

such that the following properties are satisfied:

a) Naturality: For a map $F : X \longrightarrow Y$ we have the equation

$$F_* \beta_n^X = (H_n F)^\# \beta_n^Y \quad \text{in} \quad \Gamma^b_{n-1}(H_n X; Y).$$

b) $\mu(\beta_n^X) = b_n^X \in \operatorname{Hom}(H_n X, \Gamma_{n-1} X)$.

c) $\{\pi_n(X)\} = \Delta^{-1} i^\#(\beta_{n-1}^X)$, where $i : \ker b_n^X \subset H_n X$ denotes the inclusion.

Proof. See [7].   □

The general N-dimensional case does not lead to such a good algebraic model as in the four-dimensional case. However one can prove the following theorem.

3.2.2 Theorem. Let X be a 1-connected N-dimensional polyhedron. Then X is a wedge of Moore spaces if and only if all the boundary invariants β_n^X, $n \geq 3$, are 0.

Proof. [7].   □

Remark. The β_n are naturally defined, but not workable. In particular, β_3 is the "good" k_3' above and (H_*^X, β_n^X) define a sort of "functorial" homology decomposition. There exists a equivalence relation modulo which the data of (H_*^X, β_n^X) is in one-to-one correspondence with the homotopy type of X. This can be fully understood so far only for $n \leq 5$.

We already know that β_n^X is mapped to b_n^X by the map

$$\Gamma_{n-1}^b(H, X) \overset{\mu}{\twoheadrightarrow} \mathrm{Hom}(H, \Gamma_{n-1}X).$$

A program to determine the homotopy type would run as follows:

First, choose b_4 and the extension of H_3 by coker b_4 which defines β_3 and form $X_1^4 \sim (\beta_3, b_4)$.

Then compute $X_1^n \sim (\beta_3, b_4, \dots, \dots, \beta_{n-1}, b_n)$ keeping track of equivalence of homotopy types.

3.4 Suspension of a 4-dimensional complex. There is a map of exact sequences of Whitehead

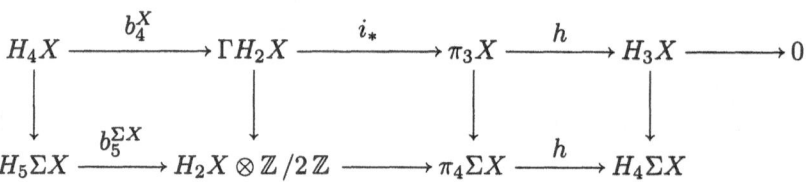

that determine the homotopy type of the suspension.

Proof. In fact for suspensions of 1-connected 4-dimensional CW-complexes (more generally for $(n-1)$-connected $(n+2)$-dimensional CW-complexes) J.H.C. Whitehead's theory works with minor modifications from the case of 1-connected 4-dimensional CW-complexes which we sketched above; see [51].

Remark. $C_*(\Omega X_1^4)$ has a similar structure: the same theory holds and the boundary invariants map from spaces to chain algebras.

Bibliography

[1] Adams, J.F. & Hilton P.J.: On the chain algebra of a loop space, *Comment. M. Helv.* **30** (1956), 305–330.

[2] Andrews, P. & Arkowitz, M.: Sullivan's minimal models and higher order Whitehead product, *Can. J. of Math.* **30-5** (1978), 961–982.

[3] Anick, D.J.: Hopf algebras up to homotopy, *J. Amer. Math. Soc.* **2**, n°3 (1989), 417–453.

[4] Anick, D.: Tame homotopy via Adams-Hilton models. Preprint.

[5] Barratt, M.G.: Track groups I & II, *Proc. London Math. Soc.* **5** (1955), 71–106, 285–329.

[6] Baues, H.J.: Obstruction Theory, *Lecture Notes in Math.* **628**, Springer-Verlag, Berlin 1977.

[7] Baues, H.: On homotopy classification problems of J.H.C. Whitehead, Algebraic Topology, Göttingen 1984, Proceedings, *Lecture Notes in Math.* **1172**, Springer-Verlag, Berlin 1985.

[8] Baues, H.J.: Algebraic homotopy, Cambridge University Press 1989.

[9] Baues,H.: Combinatorial homotopy and 4-dimensional complexes, Walter de Gruyter, Berlin 1991.

[10] Baues, H.J., Félix, Y. & Thomas, J.-C.: The Whitehead Γ-functor for chain algebras, *J. of Algebra.* To appear.

[11] Brown, E.H. & Copeland, A.H.: A homology analogue of Postnikov systems, *Michigan Math. J.* **6** (1959), 313–330.

[12] Cartan, H.: Sur les groupes d'Eilenberg-Mac Lane II, *Proc. Nat. Acad. Sci. U.S.A.* **40** (1954), 704–707.

[13] Dieck, tom T., Kamps K.H. & Puppe, D.: Homotopietheorie, *Lecture Notes in Math.* **157**, Springer-Verlag, Berlin 1970.

[14] Dold, A.: Partitions of unity in the theory of fibrations, *Ann. of Math.* **78** (1963), 223–255.

[15] Dold, A.: Halbexakte Funktoren, *Lecture Notes in Math.* **12**, Springer-Verlag, Berlin 1966.

[16] Dold, A., Thom, R.: Quasifaserungen und endliche symmetrische Produkte, *Ann. of Math.* (2) **67** (1958), 239–281.

[17] Eckmann, B., Hilton, P.: Groupes d'homotopie et dualité, *C. R. Acad. Sci. Paris* **246** (1958) 2444–2447, 2555–2558, 2991–2993.

[18] Eilenberg, S. & Mac Lane, S.: On the groups $H(\pi, n)$ I, *Ann. of Math.* (2) **58** (1953), 55–106.

[19] Eilenberg, S. & Mac Lane, S.: On the groups $H(\pi, n)$ II. Methods of computation, *Ann. of Math.* (2) **60** (1954), 49–139.

[20] Félix, Y.: La dichotomie elliptique-hyperbolique en homotopie rationnelle, *Astérisque* **176**, 1989.

[21] Félix, Y. and Halperin, S.: Rational LS-category and its applications, *Trans. A.M.S.* **273** (1982), 1–37.

[22] Félix, Y., Halperin, S., Jakobson, C., Löfwall, C. & Thomas, J.-C.: The radical of the homotopy Lie algebra, *Amer. J. of Math.* **110**, n. 2 (1988), 301-322.

[23] Félix, Y. Halperin, S. & Thomas, J.-C.: The homotopy Lie algebra for finite complexes, *Publi. I.H.E.S.* **56** (1983), 89–96.

[24] Félix, Y. Halperin, S. & Thomas, J.-C.: Engel elements in the homotopy Lie algebra, *J. of Algebra* **144** (1991), 67–78.

[25] Félix, Y. & Lemaire, J.-M.: On the mapping theorem for LS-category, *Topology* **24** (1985), 41–43.

[26] Félix, Y. & Lemaire, J.-M.: On the homology of two-level differential algebras, *Bull. Soc. Math. Belg.* 1, Ser. B, **44** (1992), 26–33.

[27] Félix, Y., and Tanré, D.: Sur l'homologie de l'espace des lacets d'une variété compacte. To appear.

[28] Freyd, P.: Homotopy is not concrete, The Steenrod Algebra and its Applications (Proc. Conf. to Celebrate N. E. Steenrod Sixtieth Birthday, Batelle Memorial Inst. Colombus, Ohio, 1970), *Lecture Notes in Math.* **168**, Springer-Verlag, Berlin (1970), 25–39.

[29] Gabriel, P., & Zisman, M.: Calculus of fractions and homotopy theory, *Ergebnisse der Mathematik und ihrer Grenzgebiete* **35**, Springer-Verlag, Berlin 1967.

[30] Halperin, S.: Lecture on minimal models, *Memoires S.M.F.*, Nouvelle Série **9–10**, 1983.

[31] Halperin, S.: Torsion gaps in the homotopy of finite complexes, *Topology* **27**, n°3 (1988), 367–375.

[32] Halperin, S. & Lemaire, J.-M.: Suites inertes dans les algèbres de Lie graduées, *Math. Scand.* **61** (1987), 39–67.

[33] Hess, K.P.: A proof of Ganea's conjecture for rational spaces, *Topology* **30**, n°2, (1991), 205–214.

[34] Hilton, P.: Homotopy Theory and Duality, Gordon and Breach, New York 1965.

[35] Hochschild, G & Serre, J.-P.: Cohomology of group extensions, *Trans. Amer. Math. Soc.*, **74** (1953), 110–134.

[36] James, I.M.: Reduced product spaces, *Ann. of Math* **62** (1955), 170–197.

[37] Jessup, B.: Rational approximation to L-S category and a conjecture of Ganea, *Trans. Amer. Math. Soc.* **317**, n°2 (1990), 655–660.

[38] Lemaire, J.-M.: Algèbres connexes et homologie des espaces de lacets, *Lecture Notes in Math.* **422**, Springer-Verlag, Berlin 1974.

[39] Lemaire, J.-M.: Lusternik-Schnirelmann category: An introduction, Nordic Summer School on Relations between Algebraic Topology and Local Rings 1983 Stockholm, *Lectures Notes in Math.* **1183**, Springer-Verlag, Berlin (1986), 259–276.

[40] Milnor, J.W. & Moore, J.C.: On the structure of Hopf algebras, *Ann. of Math.* **81** (1965), 211–264.

[41] Neisendorfer, J.A.: Ph. D. Thesis, Princeton University 1974.

[42] Puppe, D.: Homotopiemengen und ihre induzierte Abbildungen I, *Math. Z.* **69** (1958), 299–344.

[43] Quillen, D.G.: Homotopical algebra, *Lecture Notes in Mathematics* **43**, Springer-Verlag, Berlin 1967.
[44] Quillen, D.: Rational Homotopy theory, *Ann. of Math.* **90** (1969), 205–295.
[45] Samelson, H.: Groups and spaces of loops, *Comment. Math. Helv.* **28** (1954), 278–287.
[46] Serre, J.P.: Groupes d'homotopie et classes de groupes abéliens, *Ann. of Math.* **58** (1953), 258–294.
[47] Spanier, E.H.: Algebraic Topology, McGraw Hill, New-York 1966.
[48] Sullivan, D.: Infinitesimal computation in topology, *Publ. I.H.E.S.* **47** (1977), 269–331.
[49] Switzer, R.M.: Algebraic Topology — Homotopy and Homology, Springer-Verlag, Berlin 1975.
[50] Tanré, D.: Homotopie rationnelle: Modèles de Chen, Quillen, Sullivan, *Lecture Notes in Math.* **1025**, Springer-Verlag, Berlin 1983.
[51] Whitehead, J.H.C.: A certain exact sequence, *Ann. of Math.* **52** (1950), 51–110.
[52] Whitehead, G.: Elements of Homotopy Theory, Springer-Verlag, Berlin 1978.

Index

DMV Seminar
Workshops, edited by the German Mathematics Society

The workshops organized by the Gesellschaft für mathematische Forschung in cooperation with the Deutsche Mathematiker–Vereinigung (German Mathematics Society) are primarily intended to introduce students and young mathematicians to current fields of research. By means of these well-organized seminars, scientists from other fields will also be introduced to new mathematical ideas. The publication of these workshops proceedings in the **DMV-Seminar** series will make the material available to an ever larger audience.